The first book devoted exclusiv
and its use in gemstone identi₁

Until 1933 the refractometer and the microsc⟨...⟩ in gem testing. Then B.W. Anderson introduc⟨...⟩ch quickly became the third vital instrument for ⟨...⟩ ⟨...⟩mmologist. The publication of Anderson's papers in the mid-1950s alerted gem laboratories worldwide to the importance of this instrument. Today its use is universal.

Anderson and his lifelong colleague, Payne, both of the world's first full-time gemmological laboratory, were the pioneers in recognizing the vast potential of the spectroscope for this purpose, and *The Spectroscope and Gemmology,* as the extensively edited, updated and enlarged version of Anderson's original papers, provides a vital textbook wherever gemmology is used or taught. The distinguished gemmologist R. Keith Mitchell has undertaken the editing and has added chapters on the Pye spectrophotometer and on lasers. He has also reproduced new line drawings of spectra in three aspects to accommodate the prism instrument in both the British version and in that favoured by American gemmologists, and in another version as seen through the diffraction grating instrument.

Written for gemmologists everywhere and incorporating the latest developments, this book will prove indispensable to all engaged in gemstone identification.

Also Available

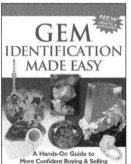

Gem Identification Made Easy, 5th Ed.
A Hands-On Guide to More Confident Buying & Selling
by Antoinette Matlins, PG, FGA, and Antonio C. Bonanno, FGA, ASA, MGA

Revised and expanded edition of the first and only book of its kind. Covers the latest gems, synthetics, treatments and instruments. Shows how to identify diamonds, colored gemstones and pearls, and separate them from fakes and look-alikes.

6 x 9, 400 pp, over 150 full-color and b/w photos and illus., index
HC, 978-0-943763-90-3

Jewelry & Gems: The Buying Guide, 7th Ed.
How to Buy Diamonds, Pearls, Colored Gemstones, Gold & Jewelry with Confidence and Knowledge
by Antoinette Matlins, PG, FGA
and Antonio C. Bonanno, FGA, ASA, MGA

A new edition of the "unofficial bible" for jewelry and gem buyers—completely updated, expanded and revised. This authoritative guidebook explains how to buy diamonds, pearls, precious and other popular colored gems, gold and jewelry with confidence and knowledge.

6 x 9, 352 pp, 16 pp in full color, over 200 full-color and b/w photos and illus., index
Quality PB Original, 978-0-943763-71-2

About the Editor

R. Keith Mitchell, FGA, Tully medallist and vice president of the Gemmological Association of Great Britain, was instructor to the G.A. courses for more than twenty years, and the author of many papers on gemmology.

Also Available

Diamonds, 3rd Ed.
The Antoinette Matlins Buying Guide—How to Select, Buy, Care for & Enjoy Diamonds with Confidence and Knowledge
by Antoinette Matlins, PG, FGA

All the information you need to buy the most beautiful diamonds with confidence—from the most widely read author in the world on jewelry and gems. This comprehensive guide is the "unofficial bible" for all diamond buyers who want to get the most for their money.

6 x 9, 240 pp, 12 pp in full color, over 150 full-color and b/w photos and illus., index
Quality PB Original, 978-0-943763-73-6

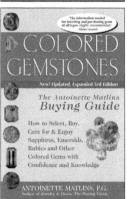

Colored Gemstones, 3rd Ed.
The Antoinette Matlins Buying Guide—How to Select, Buy, Care for & Enjoy Sapphires, Emeralds, Rubies and Other Colored Gems with Confidence and Knowledge
by Antoinette Matlins, PG, FGA

A source of expert guidance, whether buying for personal pleasure or for investment. This practical and easy-to-understand guide explains *in detail* everything you need to know to buy and care for colored gemstones with confidence and knowledge.

6 x 9, 256 pp, 24 pp in full color, over 200 full-color and b/w photos and illus., index
Quality PB Original, 978-0-943763-72-9

Illustrated Guide to Jewelry Appraising, 3rd Ed.
Antique, Period & Modern
by Anna M. Miller, GG, RMV

Completely updated and expanded, this comprehensive, step-by-step guide to jewelry identification and appraisal is indispensable for both beginners and professional appraisers. Describes in detail virtually every style of antique, period and modern jewelry, and features photos and beautiful illustrations of each style, from Georgian and Victorian to modern, "retro" and ethnic jewelry art.

8½ x 11, 216 pp, over 150 b/w photos and illus., index
HC, 978-0-943763-42-2

GemStone Press
Sunset Farm Offices, Route 4, P.O. Box 237
Woodstock, VT 05091
Tel: (802) 457-4000 • Fax: (802) 457-4004
www.gemstonepress.com

THE
SPECTROSCOPE
AND
GEMMOLOGY

Basil Anderson & James Payne

Edited by
R. Keith Mitchell

GEMSTONE PRESS
Woodstock, Vermont

N.A.G. Press
an imprint of Robert Hale · London

The Spectroscope and Gemmology

2013 GemStone Press Quality Paperback Edition, Third Printing

All rights reserved. No part of this book may be reproduced or transmitted in any form or by any means, electronic or mechanical, including photocopying, recording, or by any information storage and retrieval system, without permission in writing from the publisher.

For information regarding permission to reprint material from this book, please mail or fax your request in writing to GemStone Press, Permissions Department, at the address / fax number listed below, or e-mail your request to permissions@gemstonepress.com.

UK Edition
ISBN 0 7198 0261 X
Robert Hale Limited, Clerkenwell House, Clerkenwell Green, London EC1R 0HT

First Published in the United States of America, 1998
First Published in Great Britain, 1998

The Library of Congress has cataloged the hardcover edition as follows:
Anderson, B. W. (Basil William), 1901–
The spectroscope and gemmology / Basil Anderson & James Payne ;
edited by R. Keith Mitchell.
p.   cm.
Includes bibliographical references and index.
ISBN-13: 978-0-943763-18-7 (hardcover)
ISBN-10: 0-943763-18-5 (hardcover)
1. Absorption spectra 2. Minerals—Spectra. 3. Precious stones—Analysis. I. Payne, James (C. James) II. Mitchell, R. Keith.
III. Title.
QE396.S65A53   1998   98-12283
553.8–dc21   CIP
GemStone Press Quality Paperback Edition ISBN-13: 978-0-943763-52-1
GemStone Press Quality Paperback Edition ISBN-10: 0-943763-52-5

The right of B.W. Anderson, C.J. Payne & R. Keith Mitchell to be identified as authors of this work has been asserted by them in accordance with the Copyright, Designs and Patents Act 1988.

10 9 8 7 6 5 4 3

Manufactured in the United States of America

GemStone Press
A Division of LongHill Partners, Inc.
Sunset Farm Offices, Route 4, P.O. Box 237
Woodstock, VT 05091
Tel: (802) 457-4000   Fax: (802) 457-4004
www.gemstonepress.com

Contents

Illustrations

often faint and less diagnostic than the multiple emission lines of the finer reds. But the position of the broad absorption close to the red is useful to distinguish spinel from pyrope

13.1   PYROPE GARNET. Again a faint pattern of chromium lines with a good broad absorption of the orange, yellow and green, further from the red than in spinel. The 505nm band of almandine is usually visible

13.2   DIDYMIUM GLASS. A useful spectrum from which to assess approximate wavelengths of other spectra

13.3   POTASSIUM PERMANGANATE SOL. Another very useful comparison spectrum for assessing wavelengths of other spectra. Only a weak solution is necessary

14.1   EMERALD. This is the absorption pattern of the ordinary ray

14.2   EMERALD. The absorption pattern of the extraordinary ray. Polaroid is needed to separate these two spectra and without it emeralds may give a combination of the two in both natural and synthetic stones

15.1   ALEXANDRITE. In the red ray the shift is towards the blue, covering the yellow-green and allowing much more red through, so the colour tends to red

15.2   ALEXANDRITE. In the gamma (green) ray the broad absorption is nearer the red, leaving the green transmitted, so the stone looks green

16.1   GREEN JADEITE. Deep green specimens will show chromium lines in the red, but the most persistent band is 437.5nm and is due to iron

16.2   'WATER JADEITE'. This transparent form of pale green jadeite may show only the 437.5nm band and one or two fainter ones due to iron

16.3   STAINED GREEN JADEITE. This has a totally anomalous broad band centred at about 660nm and a fainter one at 600nm. These are due to a vegetable dye

17.1   DEMANTOID GARNET. Fine stones may have faint chromium lines in the red, but the intense absorption at 443nm is probably the best identifying band. It may be seen as a cut-off at the violet end of the spectrum.

19.1   ALMANDINE GARNET. Three intense bands due to ferrous iron provide an instantly recognizable spectrum, traces of which can often be seen in other red garnets

19.2   LIGHT PYROPE/ALMANDINE. A paler version of the

29.1   RHODONITE. A manganese spectrum seen best in the rare transparent version of this mineral

29.2   RHODOCHROSITE. A vague spectrum when compared with rhodonite. The intense band at 410nm is not easily seen at the extreme edge of human vision

29.3   SPESSARTINE. A combination of faint ferrous iron spectrum of almandine and the manganese spectrum of spessartine. Pure spessartine is rare

29.4   RED TOURMALINE. Another manganese spectrum. The sharp line in the broadly absorbed green is the main identification

30.1   Absorption curve of COBALT GLASS

30.2   SYNTHETIC BLUE SPINEL. The three band spectrum of cobalt is another easily recognized pattern. Here the central band in the yellow is broadest and most strongly absorbed

30.3   COBALT BLUE GLASS. Another three band spectrum very similar to 30.2, but here the bands are more widely spread, although varying with the type of glass, and the left-hand one in the red is now the broadest and darkest

31.1   TURQUOISE. A spectrum obtained by reflecting light indirectly from an opaque surface. The line at 433nm should be visible but 420 may be lost in the low sensitivity of the eye to the violet wavelengths

32.1   URANYL CHLORIDE. A spectrum illustrating the similarity of the known compound of uranium with the zircon spectra 32.2 LOW ZIRCON. With the crystal lattice largely destroyed the 653nm band remains as a tell-tale indication of zircon

32.3   SRI LANKAN ZIRCON. The spectrum varies in strength and in width of lines but is instantly recognizable for its ladder-like structure. This one shows twelve lines but the number and thickness can vary

32.4   BURMA ZIRCON. The strongest zircon spectrum which can give up to 41 lines of considerable intensity. This one has 23

32.5   ANOMALOUS ZIRCON. A rare 'low' form in which the 653 line has two more lines on its red side to make a prominent group of three. Other lines may not coincide with the normal zircon spectrum. The effect seems to be due to heating after millennia of metamictization

33.1   YAG DOPED WITH NEODYMIUM. A most beautiful

absorption of about 44 lines obtained by doping the near perfect lattice of artificial yttrium aluminium garnet with Nd. This is the didymium spectrum par excellence

33.2 YELLOW AND GREEN APATITE. A fine example of didymium in a natural mineral which may be found in varying strength in other colours of apatite, and very faintly in some other calcium containing gems. The rare orange scheelite has the same absorption quite strongly

33.3 DEEP BLUE AQUAMARINE. An almost sapphire blue form of beryl produced from pale material by artificial irradiation gives this extraordinary absorption in its ordinary ray only, which is the coloured one in this material. A natural gem from the Maxixe mine in Brazil has a similar spectrum, but both types fade quickly in strong light

33.4 GREEN ANDALUSITE. This rare variety from Brazil has this strange spectrum of bands with sharp edges to their blue sides. Probably due to a rare earth but which one is not known

34.1 CAPE DIAMOND. The typical absorption of yellowish 'off-colour' Cape stones. The line at 415.5nm is the key one

34.2 BROWN AND GREEN DIAMOND. A different absorption pattern from that of the Cape stones, but the two types can be found in combination. The 504nm line is the key one

34.3 IRRADIATED YELLOW DIAMOND. Shares the 'Cape' line, but 694nm and 597nm are the result of irradiation

34.4 NATURAL PINK DIAMOND. A vague broad absorption centred at 563nm

34.5 IRRADIATED PINK DIAMOND. A line spectrum markedly different from that of the natural pink stone

35.1 ZINC BLENDE. Although too soft and easily cleaved to make a usable gem, blende is beautiful when skilfully cut, and has a spectrum reminiscent of low zircon, although its visual appearance and single refraction are quite different

38.1 Fluorescence spectrum of WHITE ZIRCON

38.2 Auto-photograph of FLUORSPAR crystal exhibiting ultra-violet phosphorescence

Preface

The forty articles published in *The Gemmologist* between 1954 and 1957, which are the original source of this book, covered essentially the observations and researches into gem mineral absorption by B.W. Anderson and C.J. Payne over a period of more than twenty years.

Apart from the war period from 1939–45, when James Payne served with the Royal Artillery, these researches and investigations were a collaboration between the two young scientists, each checking and counter-checking the other's observations.

The original papers were, however, written entirely by Basil Anderson who, with characteristic generosity, included James as co-author of the large section which reported their joint research, although the latter had not contributed to the writing. Anderson was a fair-minded man and believed in giving credit where it was due.

Their research was the first to apply absorption spectroscopy to gemmology, and as far back as 1933 Anderson had introduced the method to the class at Chelsea Polytechnic as a valuable new testing procedure. Today, despite the electronic revolution of the past thirty years, it is still entirely valid and the facts as published in the 1950s remain basic to present-day practical gem identification by the ordinary gemmologist.

There have, of course, been advances and further discoveries in the subject since then and I have added relevant information, based on personal observation or culled from other sources, that has become available in later years. From this it may be seen that gem spectroscopy is perhaps more complex than was at first thought.

While recent refinements of technique may well be outside the facilities and indeed the abilities of the gem-trade gemmologist, this serves only to underline the need to retain the 'quick squint down the spectroscope' which can tell the trained gemmologist so much.

A few years back, both Anderson and I lamented the passing of the Ångström unit of wavelength measurement, for this was devised in 1868 by A.J. Ångström, a brilliant spectroscopist and very eminent Swedish scientist, and does not deserve to be relegated to historical oblivion in this way. However, the International Unit for such measurements is now officially the nanometer (1nm = 10 Ångströms) and the original text has perforce been brought into line with this ruling. This decision is mine alone for, sadly, Basil Anderson is no longer with us to make it for himself.

If any reader wishes to revert to the older measurement he need only add a zero to the nanometer wavelength value; or move the decimal point one place to the left to convert from Ångström units to nanometers.

The accuracy of the Anderson–Payne measurement of wavelengths is in general very reliable, and the original papers described how these were checked carefully against known and very accurately measured emission lines from incandescent spectra or against the Fraunhofer lines in the solar spectrum. To avoid over-long and repetitive accounts some of these sections have now been omitted. The reader will understand that wavelengths are as accurate as they could be, given the limitations of the instruments used, which really means that they were carefully checked at all times and entirely reliable within the limits of the hand-held spectroscope.

The original articles were illustrated by drawings of spectra which had been made by Mr T.H. Smith at Anderson's request, a fact which was mentioned in most of the original papers. Unfortunately only about half of Mr Smith's drawings have survived, and those scarcely in a condition to allow new plates to be made. So I have found it necessary to redraw these figures and while doing so have given them more emphasis and, hopefully, greater realism. Some of the originals did not conform precisely to Anderson's text, possibly because Mr Smith was working on his own in Hexham, perhaps with a different type of instrument, while Anderson was down in London. I have therefore done what I can to bring the illustrations more into harmony with Payne and Anderson's careful observations.

The opportunity has also been taken to print three versions of each spectrum instead of one. The first one in each set is the normal 'red on the left' prism version used in the original papers, and in most books and classes in Europe; the third one is the 'red on the right' prism version used by the Gemological Institute of America (G.I.A.); while the middle one is again the European version but as seen through an OPL diffraction-grating spectroscope, which seems now to be favoured by the Gemmological Association for their correspondence courses. I am assured that the G.I.A. have no intention to switch to this type of instrument in the near future, so there is no need for a reversed version of that spectrum to anticipate such an eventuality.

Minor adjustments have been made to the text in various places in order to bring the work up to date and to remove facts and values which no longer apply. A few sections have either been shortened, rewritten by me, or in one or two cases eliminated. Such adjustments to the text are intended to clarify and not to confuse. I hope I have achieved this aim.

The chapter on the quartz spectrograph, an instrument now almost completely superseded in modern laboratories, and that on emission spectroscopy, which again uses this instrument, have been retained more or less as originally written. This is done for the sake of completeness since they are part of a historical review of spectroscopy.

The instrument which now replaces the quartz spectrograph is the spectrophotometer, which has been given a new chapter to itself. This is a costly instrument and unlikely to come within the range of the average trade gemmologist. But it is part of the whole picture of spectroscopy and needs to be included.

The advent of artificial coloration of diamonds by irradiation has brought about some changes in our knowledge of the absorption of that gem, and has necessitated rewriting the chapter dealing with it. I have based this, in part, on a later article by Basil Anderson, so it may be regarded as being inspired by him, even if written by me. Further inspiration came from papers on diamond absorption by Dr A.T. Collins, one of the world's foremost experts on that subject. I acknowledge my indebtedness and hope that I have done justice to both gentlemen.

Apart from these changes and additions the book as a whole is essentially an account of the Anderson–Payne collaboration in the long research of this subject. For that reason it still gives rather more detail than might be needed in an ordinary testing situation. This is because Anderson was rightly concerned to record their prior claim to the initiation of the use of absorption spectroscopy in gem testing, and so felt constrained to record everything seen.

It is perhaps worth mentioning that absorption spectroscopy was in general use in British gemmology many years before it was adopted in some overseas gem testing laboratories. Indeed it was not until the last of Anderson's forty articles had been published that the subject was taken seriously by some of them – twelve years after it was introduced in the Chelsea classes. This time lag was in no way the fault of Anderson and Payne, for they had throughout been most open and generous in drawing attention to their work.

R.K.M

Part 1   The History and Development of Spectroscopy

1 'The Celebrated Phaenomena of Colours'

Although in its essentials the spectroscope consists merely of a transparent prism, or a finely ruled grating, it is still one of the most powerful investigative instruments known to science. Not only does it function as an incredibly delicate tool for chemical analysis, but it also provides a key to the electronic structure of the atom, to the configuration of organic molecules, and to the composition and motions of the stars.

In all questions relating to colour the spectroscope is the final arbiter, while its range extends far beyond visible light, further than the infra-red and the ultra-violet regions of the vast gamut of electromagnetic waves now known to us.

In the long and complicated history of the instrument, we find what so often happens in science – an initial discovery by a man of genius, ahead of his time, followed by a long period of gestation before the world seemed ready to profit by the implications of this discovery, or to make further advances. Then at last a sudden surge forward, in which workers of many nations made almost simultaneous contributions, enabling the seed sown so long before to flourish and bear fruit.

The man of genius in this case (as in so many others) was the great Isaac Newton; and no less than 150 years had to pass before further advances began to be made. Newton, in 1666, was in his twenty-fourth year, having been born on Christmas Day 1642. He had just taken his degree at Trinity College, Cambridge, and in his own words 'was in the prime of my age for inventions, and minded Mathematics and Philosophy more than at any time since.' Early in 1666, then, Newton bought a prism of glass at Stourbridge Fair, intending to try therewith 'the celebrated phaenomena of colours.' Stourbridge Fair was a famous market held on Midsummer

19

Common, near Cambridge, just off the Newmarket road. It was the origin of Bunyan's *Vanity Fair* and had a history going back to the Middle Ages. John Bunyan described a dream location probably based on the real Stourbridge, or Stowbridge, Fair. Thackeray and a later Society magazine borrowed the *Vanity Fair* title. Merchants came from all over Europe and goods reached King's Lynn by boat and thence were brought up the Ouse, so that the prism bought by Newton might have come from anywhere in Europe. In 1667 there is a record in one of his notebooks that he purchased three more prisms in London for £3; and it may have been with these that he carried out his famous experiments. His work was more than once interrupted by the spread of the Plague causing the university to close, and it was not until 1671 that a record of his observations was sent to be published in the Philosophical Transactions of the Royal Society, of which later he was to be President.

Newton's experiments

In seeking to solve the reason why, when sunlight passed through a prism of glass, colours akin to those of the rainbow were produced, Newton darkened his window with wooden shutters in which he made a quarter-inch (6.385 mm) hole to admit a narrow beam of the sun's rays. In the path of a ray of sunlight entering through this small aperture he 'placed a glass prism of refracting angle 63° 12', and observed a patch of rainbow colours, elongated to thirteen inches (33 cm), on the opposite wall. The reason for this elongation puzzled him, and he had first to eliminate by experiment several possible reasons for its appearance. The crucial experiment came when he directed his coloured rays one by one on to a hole in a board placed within the room, behind which was placed a *second* prism, and discovered that each colour was refracted to a different degree by his second prism, but that the *colour remained as before*. He thus proved what had hitherto been unsuspected, that white light from the sun was a *mixed* light consisting of rays of different 'refrangibility', and that to each degree of refrangibility there corresponded a pure spectrum colour.

A drawing showing the general arrangement for Newton's experiment (though in a Victorian setting) is reproduced in Fig. 1.1. He performed other experiments with his prisms, but those described provide the key to spectrum analysis. Although in later trials Newton used a fairly narrow slit as his light source in order to get less overlapping of his colours, he failed to observe the narrow dark lines which cross the continuous spectrum of the sun. These were actually first seen by Wollaston in 1801, when he used a crevice only 1/20th inch (1.3 mm) broad to admit the sun's rays and, standing 3–3.6 metres from this narrow source, held a prism close to his eye. The reason for Newton's failure (if one can call it a failure) in this particular probably lay in the poor quality of the prisms available to him. The hostile reception given to his theory by some eminent contemporaries (notably Hooke), involving him in controversies of which he became heartily sickened, and also his abounding interest in other branches of physics and mathematics, which were soon engaging his attention, were among the causes which prevented Newton proceeding to further work on the solar spectrum. He had, after all, achieved what he set out to do – that is to prove to his own satisfaction the reason for the appearance of the spectrum colours when light was passed through a prism, and the fundamental nature of colour itself.

Fig. 1.1 Newton's experiment proving that white light consists of mixed light of spectrum colours

The Fraunhofer lines

No further progress beyond the results achieved by Newton took place until the beginning of the nineteenth century, when the narrow lines crossing the sun's spectrum were first seen and studied. The next fifty years were to be spent in elucidating the mystery of their formation. The great name here is that of Joseph von Fraunhofer (1787–1826), a German optician and physicist of great skill and resource who (independently of Wollaston, who was actually the first to see them) not only observed the lines but mapped over 300 of them with painstaking accuracy, and later, even measured the wavelength of the principal members. He named these A, B, C, etc.; convenient labels which we still use to this day. All this, and much more, he achieved in a short lifespan of thirty-nine years. It is entirely appropriate that the dark solar lines should be known always as the Fraunhofer lines in recognition of his intimate connection with them.

Fraunhofer, like Newton, was interested in the dispersion of different types of glass from the point of view of making achromatic lenses. He had realized the need of establishing fixed points in the spectrum more clearly and exactly than was possible on the basis of colour only. Therefore he felt an added delight when he first saw the hundreds of narrow dark lines crossing the spectrum of the sun which, whatever prism he used, always had the same relative position to their coloured background. He realized that he had here, in sunlight itself, all the fixed lines he needed at his disposal. To this day, Fraunhofer lines are used as convenient intervals for dispersion measurements; gemmologists use the long range from B to G, while manufacturers of optical glass, and opticians generally, use the more restricted interval between C and F.

One of Fraunhofer's drawings of the solar spectrum is shown in Fig. 1.2.

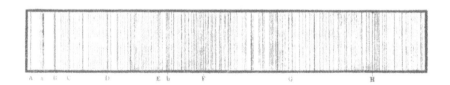

Fig. 1.2 One of Joseph von Fraunhofer's drawings of the solar spectrum. Fraunhofer eventually mapped over 300 lines in the sun's spectrum

Fraunhofer used, for his observations, a small slit 24 feet (7.3 m) away from his prism, which was mounted on a theodolite, through the telescope of which he viewed the spectrum. The prism used was an exceptionally fine one, of his own making. Under these conditions the dark lines could be very clearly seen. The paper published by Fraunhofer in 1817 records, perhaps, the greatest advances in method of any paper on the subject before or since. In it, he describes the first spectroscope, and its use for the first observation of emission lines in flames (e.g., the yellow doublet later known to be due to sodium); the first diffraction grating and the first measurement of the lines of the solar spectrum by its means; the first observation of the spectrum of a star. If Newton may be said to have found the key to the great gate leading to a knowledge of the constitution of the stars and of the atoms, Fraunhofer was the one who cleaned, inserted and turned the key. The great gate, though with much preliminary creaking, had started slowly to open.

A few more words about Fraunhofer before continuing the story of the spectroscope. As already mentioned, he was the first to develop the theory of diffraction and apply it as a practical means of measuring the wavelength of light. His earliest gratings were made by winding fine silver wire round a brass frame, taking extreme care, of course, that the spaces between the wires were constant: the minimum grating space possible for him by this method was 0.0528 mm. Now the formula for the production of spectra from a grating is $n\lambda = d\mathrm{Sin}\theta$, where n is the 'order' of the spectrum, d the grating spacing, and θ the angular deflection from the undeviated beam of light. A little thought will show the reader that this means that the first order spectra from his gratings were deflected less than a degree, giving, one would imagine, no scope for accurate measurement. Yet he obtained with ten different gratings results which varied only from 0.0005882 to 0.0005894 mm for the mean wavelength of the sodium lines – results which are remarkably near the 0.0005893 mm (i.e., 589.3nm) of modern measurement. Later, he improved on this by ruling gratings with smaller spacings on flat glass plates.

The above is a very inadequate survey of Fraunhofer's work on the spectrum, but at least it may serve to correct a gross underestimation of his work in which he was visualized merely as a clever optician who had the good fortune to stumble on the dark lines of the solar spectrum, and by painstaking labour to map and label them. In actual fact, as an experimental physicist he must be numbered among the very great.

The next really outstanding names in the history of the spectroscope are those of Bunsen and Kirchhoff; but the forty years' interval between Fraunhofer's work and theirs was full of experimental observations and advances, and it is difficult to construct a clear untangled story of what happened or to award credits to each worker who made an important contribution.

This was, in fact, the time of the sudden surge forward mentioned at the beginning of this chapter, and scientists all over Europe were making discoveries almost simultaneously and independently of one another. All that one can do here is to mention a few of those who made the most important and original contributions to the subject during this formative period.

In a paper published in 1832, Brewster suggested that the dark lines in the solar spectrum were due to absorption of light in vapours surrounding the sun, and in the following year he discovered the dark line absorption spectrum shown by nitrous acid vapour, and found that many of these lines, produced in the laboratory, coincided with lines in the solar spectrum. Finding Fraunhofer's map of the spectrum incomplete in some particulars, he was led to undertake the labour of constructing a new map of his own, on a scale 17 feet (5.18 m) in length. Brewster used fine prisms of rocksalt, oil of cassia, and of glass, and was able to insert 2,000 lines in his chart compared with 354 in Fraunhofer's drawing.

In 1834, W.H. Fox Talbot, whom we remember chiefly as a pioneer in photography, found that he could distinguish the red flames given by lithium and strontium salts by observing them through a prism and noting the very different bright-line spectrum given by each.

Professor Zantedeschi of Padua began work on the solar spectrum in 1846, and he introduced a condensing lens between the slit and the prism, placing the slit at the principal focus of the lens and thus rendering the light parallel without the awkward necessity of having the slit at a distance of many feet from the prism. Zantedeschi also used another lens beyond the prism in order to project his spectrum. This ingenious use of a collimating lens, as it is now called, between slit and prism is now, of course, a universal practice. It was adopted independently by other workers at about the same time – for instance, by Swan, when investigating the refraction of Iceland spar, and by Sims, a well-known optician of the day.

In 1849, Foucault must have been very near the crucial discovery made thirty years later by Kirchhoff, when he noticed that the D lines of the solar spectrum were made darker if the sunlight was first passed through the flame in an electric arc, suggesting that there was something in the arc which could absorb light of that wavelength – whereas, if the sunlight were cut off, the arc emitted bright lines in this same position.

In 1852, Stokes, in his lectures at Cambridge, showed that he understood the principles on which spectrochemical analysis is based, for which Kirchhoff received credit for making clear seven or eight years later.

2   The Yellow Lines of Sodium

The famous yellow lines of sodium – seen in reverse as the dark D lines of the Fraunhofer spectrum – which played so important a part in the story of the spectroscope, were actually for many years a stumbling-block to further progress. In other cases where bright lines were seen in the spectrum of flames, these could be ascribed to the glowing vapours of definite substances such as lithia, strontia, or potash. But this prominent pair of yellow lines seemed to be ubiquitous, and associated with the burning of *all* substances. For some years after Foucault's observations they had been the subject of intensive enquiry. By 1852 Stokes had succeeded in showing that they were absent from the spectrum of a candle flame if the wick had been skilfully snuffed, or from the flame of pure alcohol burned in a scrupulously cleaned watch-glass. Also, since they were most strongly developed when common salt and the other compounds of sodium were introduced into flames, it became more and more probable that they were due to glowing vapours of this element, and that the almost universal presence of sodium and the very powerful development of the lines when the merest trace was present were sufficient to explain their constant appearance. It was left for Bunsen and Kirchhoff to put this beyond dispute.

The work of Bunsen and Kirchhoff

The famous German chemist Robert Wilhelm Bunsen (1811–99) was professor at Heidelberg for no fewer than thirty-eight years. He was followed to Heidelberg by a brilliant young physicist, Gustav Robert Kirchhoff (1828–87), who was said by Bunsen's friend Roscoe to have been his greatest discovery of that period. In the late 1850s, the two collaborated very successfully in developing a convenient form of spectroscope and in establishing the origin of the bright lines in the glowing vapour of sodium and other alkali metals. They estimated that the introduction of one 200-millionth of a grain of salt into a hot flame was sufficient to produce the yellow sodium light.

Independently of other workers, they proved that the bright sodium lines were identical in position with the dark D lines of the solar spectrum. When they tried to observe the two phenomena simultaneously by passing direct sunlight through an alcohol flame containing sodium on to the slit of the spectroscope they were intensely puzzled (not being acquainted with Foucault's earlier researches) to find that, instead of seeing *bright* lines from their sodium flame, only the dark D lines were visible, and that these were much darker and thicker than before. In a famous letter to Roscoe at that time, Bunsen described how Kirchhoff, after a solid day and night of thought and experiment, succeeded in solving the problem and in producing the dark D lines artificially in his laboratory.

Fig. 2.1 Robert Bunsen. From a photograph taken on Christmas Day, 1891

He did this by passing the light from white-hot lime (which yields an intense continuous spectrum) through an alcohol sodium flame, and saw the dark D lines appear exactly as in the spectrum of the sun. Inevitably the conclusion came: there is sodium in the sun! As the result of this work, Kirchhoff was led to enunciate the following important law: that the intensity of the rays of a given wavelength which are emitted by bodies at a given temperature is proportional to the absorptive powers of the body for rays of the wavelength in question. It should be mentioned that Stokes had already explained, by the theory of 'resonance' the likelihood of both absorption and emission of light of the same wavelength by a given substance, but since his theory was propounded verbally and not published, he made no claim to priority, indeed, he had the generosity to hail Kirchhoff's discoveries as a great step forward.

Kirchhoff extended his experiments with lime-light to a flame containing lithium, and had the satisfaction of seeing dark lines due to that element, which had not been seen in the solar spectrum. Thus he succeeded in producing *new* Fraunhofer lines in the laboratory, and was able to draw the reasonable conclusion that the sun contained little or no lithium. The way was open then for a study of the composition not only of the sun, but of other stars, and soon a number of enthusiastic investigators were following this new trail. Sir William Huggins (1824–1910) and Sir Norman Lockyer (1836–1920) were among the most persistent and successful in this branch of astronomy.

It was only natural that Bunsen and Kirchhoff should spend every moment they could spare from routine duties in exploring the powers of their new spectroscope. In its simplest form this consisted of a prism mounted on a table around which were fixed three horizontal tubes. One contained the slit and collimating lens, the second was the observing telescope, focused at infinity, and the third contained an arbitrary scale. This scale was illuminated by a small luminous flame, and its image could be seen reflected from the prism face, superimposed on the spectrum, by the observer when looking through the telescope. This single-prism table model spectroscope proved so well suited for the observation of flame spectra that it was in use in chemical laboratories until quite recent times. For solar spectroscopy, instruments with a larger dispersion were needed, and multi-prism spectroscopes were used, each of the chain of prisms being set in the position of minimum deviation. Just as nowadays there was a certain amount of competition as to who could build the most powerful cyclotron, there was at that period a tendency for spectroscopists to vie with one another in the number of prisms they employed.

Early in 1860, Bunsen, while examining the flame spectrum of concentrates from Dürkheim mineral water, saw, in addition to the now familiar lines of lithium, sodium, potassium, calcium and strontium, two remarkable blue lines close together, which heralded the presence of a new alkali element. He christened this *Caesium* on the basis of its blue spectrum lines. No less than 40 tons of the mineral water had to be evaporated to produce the 17 grams of caesium chloride needed for Bunsen's subsequent work on the properties of his new element and its salts. The caesium mineral pollucite (clear samples of which are coveted by gem collectors) had already been analysed in 1846 by Plattner, but he was unable to make his analyses total 100 per cent, as he mistook his caesium sulphate for a mixture of sodium and potassium sulphates. In 1861 the spectroscope led Bunsen to the discovery of yet another alkali metal which he named *Rubidium* from the spectrum lines, even deeper in the red than Fraunhofer's A line, which chiefly characterize it.

Spectrum analysis led, in other hands, to the discovery of further new elements. In 1861 Sir William Crookes was the first to observe the strong green line of *thallium* in selenium-rich residues from a sulphuric acid plant. In 1863, the discovery of *indium*, also named from the colour of its spectrum lines, was the subject of a quarrel between rival claimants. Reich isolated by purely chemical means the sulphide of what he rightly suspected to be a new element, but, being colourblind entrusted the spectroscopic examination to his assistant Richter. Shortly afterwards, Richter was unscrupulous enough to try and claim sole credit for the discovery. Later, gallium and other elements were discovered by means of their spectra, but, as can be imagined, there were also numerous ill-founded claims to the discovery of new elements by means of this fashionable and deceptively easy method by scientists in an undue hurry to acquire fame. Even so good an investigator as Sorby made too hasty an assumption of this nature when he first saw the remarkable absorption lines in zircon.

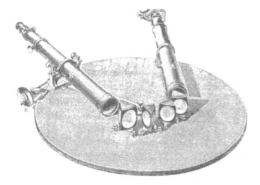

Fig. 2.2 Four-prism astronomical spectroscope made for Kirchhoff by Steinheil

Less spectacular but fundamentally important and necessary work was being carried out during this period by A.J. Ångström (1814–74) and later by Thalén at Uppsala. Kirchhoff and his pupil Hofmann had already prepared a drawing of the Fraunhofer spectrum on a vastly extended scale – the whole range from A to G occupying eight feet (2.4 m). For this he used a four-prism spectroscope (Fig. 2.2). But the positions of the lines were recorded on an arbitrary millimetre scale, and the drawing was therefore of comparatively little use to other observers using other spectroscopes. Ångström, on the other hand, used accurately ruled diffraction gratings made for him by a German optician named Norbert, on a machine of his own secret design. Unlike his predecessors, Ångström was thus enabled to draw his lines on a regular wavelength scale, using units of a ten millionth of a millimetre, which became familiar as the convenient 'Ångström unit'. In the original maps, published in 1869, the whole solar spectrum from red to deep violet was exhibited in eleven sections, and contained some 1,000 measured lines. The entire spectrum extends to about 3.35 metres. Where possible, the more prominent dark lines were also marked in – a truly formidable task (Fig. 2.3).

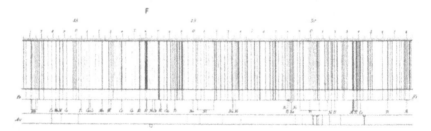

Fig. 2.3 The F section of Ångström's great map of the solar spectrum (reduced)

Ångström's measurements were accurate enough to serve spectroscopists for the remainder of the century, but even better data were provided by Rowland some two decades later. Rowland designed a machine with which he was able to rule quite exceptionally perfect six-inch (15.24 cm) concave gratings on speculum metal – an alloy of 66% copper and 33% tin which has brilliant reflecting properties. Modern experts consider some of Rowland's gratings the finest ever made by this method. More recently rulings have been made in the form of a very fine screw thread on a metal cylinder, which is then used to control a second ruling machine in such a way as to average any minute errors over many hundreds of lines. A master thread of great accuracy is thus made, from which vast numbers of pressed plastic gratings can be made. These are then silvered.

Today such gratings are 'blazed' at exact angles to produce the brightest possible spectra, and in some cases they are produced holographically.

With the aid of his gratings Rowland set a new standard of accuracy in wavelength measurement. His *New Tables of Standard Wavelengths* were published in an astronomical journal in 1893, but books of wavelength tables were not readily available until years later.

3   The Absorption Spectra of Solids

In this brief historical introduction to the spectroscope and its uses, we have so far confined our attention to the sharp, narrow dark lines seen in the solar spectrum and the equally sharp bright lines corresponding to these, which are emitted by the glowing vapours of various elements.

As a general rule, the absorption bands seen in the spectrum of light which has passed through what the physicist terms 'condensed systems', i.e. through solids or liquids, are far less narrow and well defined. Here the atoms and ions (electrically charged atoms) are no longer able to undergo unimpeded the electronic shifts which we now know to be responsible for the line spectra of glowing vapours. The movements of the outer electrons are so modified and interfered with by the forces exerted by the neighbouring atoms that only in a very few special cases can any relationship be traced between the emission spectrum of an element and absorption bands for which it may be responsible in a solid.

It should perhaps be added that the absorption spectrum even of a vapour or gas *in the cold* may bear no apparent relationship to the lines it emits at high temperature. Chlorine is an example of this. Thus it must be clearly understood at the outset that wavelength tables of emission spectra of the various elements give no clue whatsoever to the origin of the absorption bands seen in minerals. Indeed, many of the bands with which we are familiar as an aid to recognizing a certain gemstone cannot yet be ascribed with any certainty to any particular element. Others, such as those due to the presence of chromium, can be recognized with complete certainty as belonging to that element.

The first observations on the absorption spectra of solids were carried out by David Brewster as early as 1833, when he described the typical broad bands due to cobalt in specimens of blue glass.

The Absorption Spectra of Solids

As to the gem minerals, probably the earliest record occurs in Stokes' great paper on fluorescence (1852)[1] in which he makes passing mention of an absorption band in a crystal of green fluorspar from Alston Moor, which exhibited a deep blue fluorescence. 'On viewing a pure spectrum through it, there was found to be a dark band of absorption in the red. This band was narrow, and by no means intense'. In the same paper, Stokes describes the absorption and the fluorescence spectra of uranium salts and minerals, and notices the great similarity between the nature and spacing of the two sets of bands.

> After having seen both systems, I could not fail to be impressed with the conviction of a most intimate connexion between the causes of the two phenomena, unconnected as at first sight they might appear. The more I examined the compounds of uranium, the more this conviction was strengthened in my mind.

But the credit for describing absorption bands in gemstones which could be used for diagnostic purposes belongs properly to Professor A.H. (later Sir Arthur) Church, who in 1866, saw the striking absorption bands in zircon and in almandine garnet. He described his discovery in a letter to the *Intellectual Observer*, which appeared under the heading 'Microspectroscope Investigations'.[2]

> I have worked lately on the spectra of pleochroic minerals and salts. Among the minerals recently examined were fine specimens of the true zircon or jargoon, a silicate of zirconia. These gave a beautiful and most characteristic system of seven dark bands quite different from those belonging to any other substance yet examined. They are roughly sketched in the following figure. Zircons as colourless as common glass show these bands as well, perhaps better, than those possessed of colour. I incline to think that those zircons which have come from some localities show the bands better than those from others. Several Expailly specimens scarcely exhibit anything of this kind; all those from Ceylon and Norway show the bands well.
>
> From this observation I am induced to hazard the conjecture that it may be, after all, the presence of Swanberg's *norium* which determines the difference. You are aware that the orange jacinth, a variety of zircon, is very precious, and

that the essonite or cinnamon garnet is constantly sold for it. Curiously enough, the cinnamon garnet (a lime-garnet) has no conspicuous dark absorption bands at all, and so the spectroscope may be brought to bear upon the discrimination of these two stones. We have thus a much more ready process than that of taking the density of the specimens. The lime-garnet is of comparatively small value. The iron-garnet of different shades (carbuncle, almondine, [sic] etc.), gives a beautiful and very characteristic spectrum with several intensely deep absorption bands. I ought to add that the absorption bands of zircon resemble those of didymium, discovered by Gladstone, in their sharpness and in their being produced by the passage of light through a colourless medium.

Church has been quoted at some length, partly because of the importance of his discovery in the history of gemmology, but also because of the interesting sidelights it contains.

It will be noticed that Church suspected that a hypothetical new element 'norium' might be present in zircon from some localities and be the cause of the mysterious bands. Later, the well-known mineralogist, H.C. Sorby, unaware of Church's prior work, also discovered the remarkable zircon spectrum, and was led to assume the presence of a new element, which he christened 'jargonium'.[3] It is to his credit that he was later able, after much experimental work, to ascribe the bands to their true cause – the traces of uranous compounds which are almost universally present in zircon.

It is a very curious fact that Church, who had a profound and lifelong interest in precious stones, did not, after his initial exciting discovery, turn his spectroscope on to the other coloured gems, and find in some of them at least, the absorption bands which they exhibit quite clearly. It is probable that he did make some further experiments, but was unlucky in his choice of specimens, and had not a really suitable spectroscope with which to see what to us is now so obvious. One can hardly blame the low power of his light-source, since almost certainly he used the sun. Whatever the reason may be, the legend persisted until quite recent times that zircon and almandine garnet were the only two gemstones which showed characteristic absorption bands; and with a function apparently so limited it was natural that the spectroscope should not have been regarded at all seriously as a gem-testing instrument. Authors of text-books are partly to be blamed. They remained obstinately unaware of much work on the subject which had appeared in the scientific journals, finding it easier to cull their information from earlier books.

Luminescence spectra

Pioneer work in another branch of spectroscopy which should interest gemmologists was carried out by the French physicist E. Becquerel in 1859.[4] He had invented an ingenious phosphoroscope, in which light was admitted through slots cut in a spinning disc and caused to impinge on the specimen to be observed. This was viewed through a second spinning disc in which the slots were staggered in relation to those in the first. Any light seen from the specimen was therefore purely a phosphorescent emission stimulated by the incident rays which had fallen upon it a fraction of a second earlier. Becquerel analysed the phosphorescent light given out by a number of minerals and salts by means of a table spectroscope, and made careful drawings of the effects he saw – fixing the position of his bands as nearly as possible with reference to the fixed lines of the Fraunhofer spectrum. Here we have the first description of the luminescence spectra of ruby and spinel, showing the distinctive grouping of narrow lines in the red which we find so useful today. Some half-a-dozen broad emission bands are shown in the spectrum of a fluorspar, and the strikingly regular series of emission bands shown by uranium salts is also pictured. The phosphorescence of diamond is shown, but no sharp lines or narrow bands are given – only some rather vague positions of maxima and minima. This fine work of Becquerel's, more than a hundred years ago, has much to commend it to workers of the present day. The mere observation of a 'red' or a 'greenish' or 'orange-yellow' luminescence under ultra-violet light can be made much more precise and revealing if analysed with a spectroscope – though admittedly in many cases the glow is a continuous one between rather ill-defined limits.

Invisible light

Before pursuing the story of the spectroscope into the latter part of the nineteenth century, a few words should be said about the pioneers who first discovered the existence of invisible rays below the red end of the spectrum and beyond the deepest violet. This knowledge came surprisingly early; actually before the discoveries by Wollaston and by Fraunhofer of the dark lines in the solar spectrum. In 1800, William Herschel noticed that the sun's rays gave to different extents the sensation of heat when passed through different coloured glasses. This led him to place the blackened bulb of a sensitive thermometer in different parts of the spectrum from the sun, and to compare the temperature recorded with that of a similar thermometer in the shade. He found an increasing rise of temperature towards the red, and was astonished to find that the increase persisted well beyond the red, into the invisible region that we now call the 'infra-red'. The thermometer scale used is not recorded, but the rise in temperature in the violet was 2°; in the green, 3½°; in the red, 7°; while a maximum of 9° was attained well beyond the visible limit of the spectrum.

Herschel's discovery was published in the Philosophical Transactions of the Royal Society. In a brief note in Gilbert's *Annalen* early in the following year, J.W. Ritter describes how the blackening of silver chloride which Scheele had found to occur when this chemical was exposed to light of the visible spectrum was even more pronounced when exposed to invisible radiations beyond the violet, and that the extent of this invisible ultra-violet field was '*sehr gross*' (very great).

This discovery of the ultra-violet rays by Ritter was not effectively followed up for many years, largely because the photographic process had not yet been invented. In 1842, E. Becquerel projected a well-defined solar spectrum on to an iodized silver plate, which was later 'developed' by exposure to mercury, following Daguerre's process. His image showed a number of the Fraunhofer lines, and these extended well beyond the visible violet.

Professor Draper of New York produced photographs of ultra-violet spectra at about the same time, and was perhaps the first to use an adjustable slit in his spectrograph. Next, Stokes, in his great 1852 paper on fluorescence, described some prominent Fraunhofer lines beyond the violet which he was able to observe visually when a spectrum was projected on to a fluorescent quinine sulphate screen; while ten years later both Stokes and W.A. Miller published papers on ultra-violet absorption in the same volume (152) of the Philosophical Transactions of the Royal Society. Stokes was using fluorescent screens, but Miller's apparatus was a true quartz spectrograph. He recorded the lengths of the spectra on his photographs after the light had been transmitted in turn through different minerals and liquids. Careful scrutiny of his results shows that, by an extraordinary chance, two of the three diamonds he used were of the rare variety which is unusually transparent to the far ultra-violet rays, which later became known as 'Type II' diamond.

4   Other Uses of the Spectroscope

For a few years after Bunsen and Kirchhoff's discoveries of 1859–61 the amazing powers of the spectroscope as a tool for detecting the presence of elements both old and new in terrestrial matter, and in revealing to astronomers the composition of the sun and stars, had roused the enthusiasm of scientists throughout the civilized world. But before long the enthusiasm began to cool.

There were several reasons for this. There was the natural reaction when the novelty of the new scientific toy had worn off (in science, as in other human affairs, there are 'fashions' in apparatus and methods). There was also the growing realization of the complexity of the subject. Simple flame spectra of the alkali metals and of the alkaline earths were easy to recognize and study, but the maze of thousands of lines in the arc spectrum or spark spectrum of metals such as iron formed a far more daunting phenomenon. One must also remember that photography as we know it was hardly born, that electricity was not generally available at the turn of a switch, and that suitable instruments embodying a quartz optical system for photography of spectra in the ultraviolet region (in which region the most significant lines of most elements appear) were not commercially available.

The astronomers, however, were kept very busy and very happy with their new tool, which to them became second in importance only to the telescope itself. In 1842, Doppler had suggested that when a source of light was approaching towards or receding from the observer at a high velocity in the line of sight it should undergo an apparent change in frequency analogous to the well-known change in the pitch of sound when the body emitting the sound is first approaching and then receding from the listener. In the case of an express train travelling at a speed of sixty miles (96.5 km) an hour and whistling as it goes, there is a sudden drop of pitch of three semitones audible to a man standing on a station platform, at the moment when the locomotive passes the point where he is standing.

This relatively large effect with (by sidereal standards) a slow-moving object like a train is due to the low velocity of sound waves – 1,100 feet (335.3 m) per second. With light, which has a velocity of nearly 186,300 miles (300,000 km) per second, the Doppler effect could only be perceptible if the light-source were moving at enormous speed in relation to the observer. Doppler originally suggested that stars should appear of different colours according to their motions in the line of sight; but, considering that there are, so to speak, large reserves of invisible wavelengths at either end of the visible spectrum, even quite a large change in frequency could make no difference in the colour of a star, since waves moving towards the ultra-violet would be replaced by others coming in from the infra-red, or vice-versa, and the sum total of radiation from the star would be virtually the same. Fizeau, in 1848, put the theory on a sounder basis, and in consequence it is often called the Doppler–Fizeau principle. By concentrating his attention on particular lines in the spectrum of some of the brighter stars Sir William Huggins, who was one of the pioneers in stellar spectroscopy, was able to establish small but definite differences between the frequency of these lines and the frequency of the same lines when these were produced by an arc or spark in the laboratory.

Modern giant telescopes and improved photographic techniques have extended both the range and accuracy of such observations and, although there can be no doubt as to the validity of the Doppler effect, astro-physicists are now wondering whether some other cause than rapid recession can cause the extensive shift towards the red of spectrum lines which is measurable in light reaching us from the more distant nebulae. In a typical case the powerful H and K lines of ionised calcium, which in the sun's spectrum have wavelengths 396.8 and 393.4nm, appear in the nebular spectrum in the 450 region – which corresponds to a velocity of about 42,000 kilometres per second away from the earth, or approximately 1/7th the speed of light itself. Apart from the sheer magnitude of this velocity there is the puzzling fact that the more distant these heavenly objects are from the earth, the greater their velocity away from it. So regular is this relationship that the formula $V = 170 D$ (where V is the speed of recession in km per second, and D the distance in millions of light-years) is found to hold with fair exactness.

The Doppler effect actually makes itself apparent under laboratory conditions when light emitted by an incandescent vapour is analysed by the spectroscope. The atoms in such a vapour will be in violent motion, and those which happen to be approaching the observer at any instant will have a slightly different frequency for the spectrum lines they emit from those which are receding. This is noticeable as a widening of the emission lines concerned. The hotter the vapour the wider the line, if other conditions are the same.

The Doppler effect has led to many other startling astronomical discoveries such as that of the so-called 'spectroscope binaries', a numerous class of stars in which what appears to even the largest telescopes as a single body is found to consist of two stars revolving round each other. In such double stars a single bright spectrum line is seen periodically to separate into two components, light from one star shifting to the red, the other towards the violet, and then to close up again to a single line when the two stars are both in the same line of vision. This is not the place to go further into the astonishing amount of information that the spectroscope has brought to the astronomer. It was the physicist who gave to astronomy this powerful instrument, but with its aid the astronomer has in turn enriched physics by revealing the behaviour of atoms under the colossal temperatures and pressures obtaining in the stars. As Eddington pointed out, man himself stands almost exactly half-way in scale between the atoms and the stars. About 10^{27} atoms build the body of a man, and about 10^{28} human bodies would constitute enough material to build an average star.

Hartley and de Gramont

Towards the close of the nineteenth century spectrochemical analysis began to return to favour, largely thanks to the influence of a series of researches carried out by W.N. Hartley and his assistants at the University of Dublin. Hartley noticed that it was not always the strongest spectrum line of an element which is the most persistent when the substance is present in diminishing amounts, and he set to work preparing solutions of metallic salts in controlled degrees of dilution, recording the spark spectra of these by means of a quartz spectrograph. He was thus able to categorize the most persistent lines of many of the elements in order of their sensitivity, thereby laying the way open to a roughly quantitative estimate of the metals concerned if these are present in small amount.

Very similar work was undertaken during the first two decades of the present century by A. de Gramont, working at the Sorbonne in Paris. Hartley's 'persistent lines' were termed *raies sensibles* by de Gramont, who reserved the name *raies ultimes* for those which remained last of all when the quantity of the element was decreased almost to vanishing point. The *raies ultimes* of most elements lie in the ultra-violet region of the spectrum – hence the value of a quartz spectrograph in chemical analysis. Quite a number of the *raies sensibles*, however, are to be found in the visible region, and enable very small quantities of the elements concerned to be detected with quite simple apparatus. Many modern books on spectroscopy contain lists of the strongest lines of the different metals as seen in the arc and in the spark spectrum (the two are by no means identical) together with a numerical indication of their relative strength, from 1 to 10. The most persistent lines are usually marked by special symbols. The whole conception of persistent lines and *raies ultimes* has simplified enormously the business of establishing what elements are present in a sample. If the *raies ultimes* of an element are missing, it is a waste of time to search for other, less persistent lines.

Theory of spectra

While the chemists and the astronomers were busy developing their own special techniques in spectroscopy, the physicists were concerned with attempts to find some general mathematical relationship between the wavelengths or frequencies of lines emitted by any one element. The spectrum of hydrogen as exhibited in light from a discharge tube filled with the gas was one of the first to be studied. It had the advantage of simplicity – there are only four lines in the visible spectrum under these conditions. These are the Fraunhofer lines C and F at 656.3 and 486.1nm much used by opticians for dispersion measurements, and Fraunhofer's G. and h. at wavelengths 434.1 and 410.2nm. If these are drawn to scale on squared paper it is at once evident that they form a converging series, and the lines continue to get closer and closer in the ultra-violet until they reach a limit near 365nm. In addition to becoming more closely spaced towards the end of the series the intensity also becomes progressively less and it is difficult actually to observe the concluding close-packed lines.

Between 1885 and 1913 several scientists, among them Balmer, Lyman, Rydberg and Paschen, contrived mathematical formulae to show the relationship between the lines of such diminishing series. But however closely these described the relationships, such formulae were, of course, purely empirical, and it was not until the end of that period that the young Danish physicist Neils Bohr, by the application of his intuitive genius to the problem, succeeded in linking Planck's quantum theory (first promulgated in 1900) to the planetary model of the atom, which had been recently evolved by Rutherford. The atom was considered to consist of a small positively charged nucleus around which negatively charged electrons circulate. Bohr broke completely with the traditions of classical physics, which in such a system, would require the electrons to describe a spiral orbit with eventual collapse into the nucleus, emitting the while a continuously varying radiation. He boldly postulated that electrons can move in certain selected orbits without radiating, and therefore without losing energy. Only when an electron which has been forced by some outside influence to move into an outer orbit and makes a sudden jump from this orbit to one of lower energy can radiation be emitted, and the frequency of the radiation is determined by the difference in energy between the two stationary states concerned. Bohr's original theory succeeded very well in 'explaining' the frequencies actually measured in the spectrum of hydrogen. In more complex spectra it was not so successful. Disturbing factors such as the elliptical nature of the electronic orbits, electron spin and so on, had each to be taken into account to explain the discrepancies. Sommerfeld and other continental workers succeeded very well in grappling with the enormous difficulties involved in this new 'wave-mechanics', and the 'energy levels' of most of the atoms have by now been largely worked out from spectroscopic data.

The spectroscope has thus contributed very significantly to our knowledge of the structure of atoms as well as of the stars, the two extremes in the vast scale of size in the universe.

On the practical side, spectroscopy made enormous strides in the twentieth century and even forty years ago there were few industrial or research laboratories which did not have spectrographs of some kind among their essential equipment.

5 Emission Spectra

It is part of the gemmologist's code that the tests he applies to precious stones should leave the specimen unharmed in any way. Thus, faceted stones must not be scratched with hardness points, porous stones must be denied the pleasure of a bath in heavy liquid and stones susceptible to colour changes must not be heated or exposed to short-wave radiations. By the same token, chemical analysis in any form is usually considered out of the question in routine gem-testing, since even the most delicate of chemical tests does demand the sacrifice of a small portion of material from the specimen.

The position is quite different where rough stones are concerned, or where the gemmologist is engaged in research on material in his own possession; also, in the case of beads or carvings purporting to be turquoise, jade or amber, it may be legitimate to remove minute portions of material for a chemical test. M.D.S. Lewis and Robert Webster were among those who have shown the value of sensitive chemical 'spot tests' in certain cases, and other workers have turned to the spectroscope as the quickest means of detecting the metallic elements present in a small sample of mineral substance.

Unfortunately, a simple direct-vision spectroscope of the type which can be used so successfully in the study of absorption spectra has a very limited value when applied to the far more complex field of emission spectrum analysis. For most work of this kind an arc lamp is obligatory and some form of spectrograph highly desirable, so that the whole process becomes essentially a laboratory one, and for that reason will only be described here very briefly. The absorption spectra of gemstones, on the other hand, can be studied with far more modest apparatus and are in general much more diagnostic; these will accordingly be described in very full detail later.

Flame Spectra

It will be remembered from the historical survey that, apart from the sun, the earliest light-sources studied with the spectroscope were flames such as those of the alcohol lamp or the hot gas flame devised by Bunsen in his well-known 'burner', into which salts of various elements were introduced. The early discoveries of Bunsen and Kirchhoff were indeed based on a careful study of the flame spectra of the alkali metals and alkaline earths. Of the alkalis, lithium, sodium and potassium, were already known, while caesium and rubidium were found by means of their distinctive spectrum lines.

'Alkaline earths' is the name commonly used for the closely related divalent elements, calcium, strontium and barium. Salts of these elements give a more complex spectrum in the bunsen flame, in which narrow bands due to compounds of the elements are seen as well as a few sharp lines due to the metals themselves. Further, it will be remembered that thallium and indium were two rare new elements discovered later than the rare alkalis by means of their distinctive flame spectra, the names of the elements being based on the colour of the lines seen. But there the tally of flame spectra ends – at least so far as the ordinary gas-burner is concerned. With practically all other elements higher temperature flames such as the oxy-hydrogen flame, or even the exceedingly high temperature of the electric arc (3000°–6000°C) are needed to volatilize and dissociate compounds and energize the atoms sufficiently to cause them to emit spectrum lines. Where the relatively low temperature of the bunsen flame (about 1500°C) produces the well-known yellow lines of sodium, at higher temperatures further sodium lines are brought into being, extending into the ultra-violet – and for most elements the most sensitive and powerful lines are to be found in the ultra-violet region.

For flame analysis with a direct vision hand spectroscope a minute scraping of the specimen is collected on a scrupulously clean loop of platinum wire and held in a colourless Bunsen flame near the base of the outer cone. The coloured flame which then appears above the wire is examined through the spectroscope. The yellow doublet of lines due to sodium is bound to be seen, for that element is almost ubiquitous. Lines due to other elements may need practice before they can be readily recognized, and in any case these are limited to those produced by the alkali and alkali earth elements mentioned above.

A vastly more important field is opened up when the test material is added to the intense heat of a carbon arc, but here some form of wavelength measure is essential if an analysis by means of emission lines is to be attempted.

The table spectrometer

The multi-prism spectroscope of Bunsen and Kirchhoff was very successful for elementary work, but is now considered obsolete although with suitable calibration and adjustment it could still be very useful.

More recently this type of spectroscopy was carried out on the table spectrometer, which is essentially the same instrument as the reflecting goniometer with which very accurate measurements of interfacial crystal angles can be made.

Fig. 5.1 gives a simplified plan of this instrument. There is a central rotating table on which a prism is clamped. The table is surrounded by a fixed ring graduated in degrees and half-degrees. A collimating tube, with a slit at one end and a lens at the other, is attached rigidly to the graduated circle. A second tube is mobile, free to swing around the central axis of the table, and consists of a telescope with crossed hairs in the eye-piece. The rotating table and the telescope each have their own verniers which move on the fixed scale and allow measurements accurate to at least one minute of arc, that is to a sixtieth of a degree.

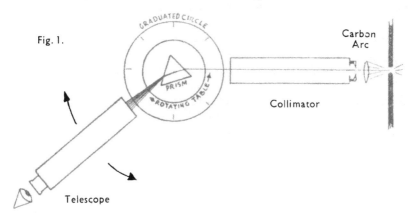

Fig. 5.1 The table spectrometer

When the telescope and collimator are correctly mounted, light passing through the slit should be seen centred on the cross-hairs of the eye-piece, with the slit in sharp focus when the telescope is adjusted to infinity.

For spectrum analysis a good quality 60° prism of flint glass with an R.I. of about 1.625 should be placed in the centre of the rotating table and turned to a position which allows an image of the collimator slit to be seen through the telescope.

A carbon arc provides the most suitable brilliant source of light for visual spectrum analysis. The arc is focused by a lens on to the slit, all of which are strictly in line. If they were not, no spectrum would be seen. Finally a black screen of metal or card is used to shield the observer's eyes from the intense direct brilliance of the arc. The ordinary carbon arc is not so dangerous as an ultra-violet arc, but to look at it can still affect visual acuity for quite a while, and it is better anyway to avoid the additional strain involved when such an intense light enters the eye. When inserting samples for analysis into the burning arc dark goggles *must* be worn.

Even if the observer is already familiar with the solar spectrum and other continuous spectra as seen through the pocket spectroscope, the arc spectrum cannot fail to impress with its beauty when viewed for the first time through the telescope of a good spectrometer. Against the background of the continuous spectrum the yellow sodium lines will be clearly recognizable, while down in the violet the serried lines of the cyanogen bands will inevitably appear. The cross-hairs of the telescope should be centred on the sodium lines, which, if correctly focused, should just be visible as a close *doublet*. The prism table, with the prism on it, should be slowly rotated in whichever direction diminishes the angle of refraction, and the image of the sodium lines followed with the telescope until a critical point is reached beyond which, no matter which way the prism is turned, the spectrum will no longer move towards a lower angle of refraction, but will start to return towards the direction of greater refraction. This critical position of *minimum deviation* for sodium light should be invariably adopted as the standard position for the prism before carrying out any measurements on spectrum lines, to ensure as far as possible that these are comparable for different experiments – provided, of course, that the same prism is used and that the optical train is truly level in each case. A prism holder or clamp is usually supplied with the instrument and the prism should be secured with this to ensure that it does not move from its correct setting if accidentally jolted. For a similar reason the prism table also should be clamped when the position of minimum deviation has been found.

The carbon arc uses mains alternating electricity adjusted by a variable resistance to about 6 amps. Ordinary best quality commercial cored carbons are suitable and may be further improved by leaching the ends to be burned in pure hydrochloric acid, followed by distilled water, to remove most impurities.

The spectrometer is calibrated in the first instance by inserting various substances into the burning arc. Sodium will invariably be present in the carbons, then lithium mica (lepidolite), chalk or

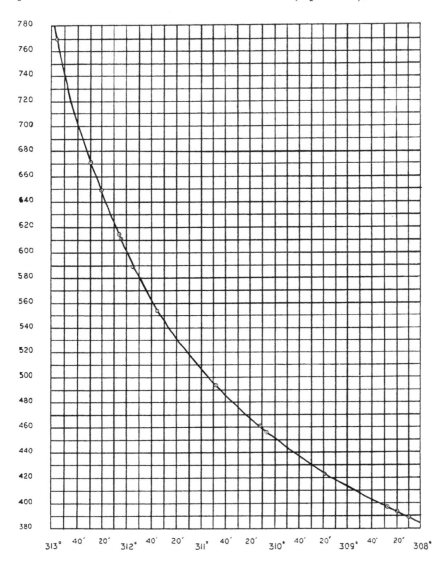

Fig. 5.2 Plotting wavelengths against angular readings on the instrument scale

47

calcite, barytes and Clerici solution are suggested to give a range of bright, clearly recognizable narrow emission lines spread through the visible region of the spectrum. A graph is then prepared plotting wavelengths against angular readings on the instrument scale; as in Fig. 5.2 in which it will be seen that the plotted values fit very neatly on to a regular descending curve. Readings taken from any other substance, using new carbons, should then fit into this graph reference curve which is standard for that prism. If another prism has to be substituted for any reason then it must be similarly calibrated before it can be used. Such calibration curves are kept for further reference.

If Fig. 5.2 is studied it will be seen that the whole visible spectrum from 700 to 400nm covers a mere 4° on the spectrometer scale. From this it is obvious that the single prism set-up cannot pretend to great accuracy, and for this reason early workers favoured a hollow glass prism filled with carbon disulphide, or cinnamic aldehyde, each far more dispersive than flint glass; or a series of prisms. A diffraction grating also gives a longer spread, even in a 1st order spectrum, and has the additional ability to space the wavelengths equally. But the brilliance is far less than that seen through a prism.

Any work on this type of instrument calls for the use of wavelength tables with which to identify the positions of lines which each element may be expected to produce. Such specialized publications are still available although those recommended in the original papers are long out of print.

6 The Quartz Spectrograph

The table spectrometer and the quartz spectrograph are research instruments which have been largely superseded by various versions of the spectrophotometer, which will be described in the next chapter. But they were each valuable instruments in their day and in an historical survey they have to be included.

The quartz spectrograph uses a prism and lenses made from quartz which, unlike glass, transmits ultra-violet wavelengths freely down to below 200nm. Ordinary photographic emulsion is also sensitive to this region and a range of the spectrum is therefore opened up which may contain the most definitive lines for some elements, which makes this form of spectrograph a much more powerful analytical tool than the simpler spectroscopes so far discussed.

A crude form of quartz spectrograph was devised by Crookes in 1861 – only a year or two after Bunsen and Kirchhoff had published details of their classic form of prism spectroscope for visual spectrum analysis. A year later, W.A. Miller took some badly focused photographs of ultra-violet spark spectra, using wet collodion plates. It was not until 1877 that another British physicist, W.N. Hartley, had constructed for him the first satisfactory quartz spectrograph, and with this he carried out the famous series of researches mentioned earlier.

Despite the work of these early pioneers, the practice of spectrographic analysis was in the doldrums towards the close of the century, and instruments capable of accurate work and so designed as to be convenient in operation were not commercially available until the early years of the present century, when the famous scientist and optician, F. Twyman, designed a series of instruments for the firm of Adam Hilger, in London. It is hardly too much to say that all quartz spectrographs subsequently made, whether by Hilger or other firms at home or abroad, were basically derived from these early designs of Twyman.

Three types in use

In accordance with the type of work to be undertaken there were, broadly speaking, three types of spectrograph in use, and these are simply and conveniently known as the 'small', the 'medium' and the 'large'. The small spectrograph was naturally the least ambitious and least expensive, and the entire visible spectrum and the ultra-violet to 200nm extended to only about 7.5 cm, which naturally resulted in a hopelessly crowded series of lines when metals such as iron, chromium or cobalt, which have very complex spectra, were present in considerable quantity in the sample to be analysed. For simpler spectra, such as those of the alkaline earths, lead, tin, magnesium, aluminium, silicon and zinc, a small instrument of this kind was capable of excellent work. Many of the gem minerals contain very little iron, and with these a skilled worker could gain a very good idea of the main metallic contents in a mineral sample.

In the medium spectrograph the total length of spectrum was about 23 cm, while in the large spectrograph the dispersion gave a spectrum length of 60 cm, and needed at least three separate exposures at different points along the spectrum to record the whole range. Later, instruments were designed so that this operation was accomplished by turning a single drum.

In all these instruments the components were essentially the same – slit, collimating lens, prism, camera lens and plate holder; but the accuracy with which these components were made and adjusted called for supreme skill, and accounts for their high cost. The slit, for example, must have its jaws not only strictly parallel at all adjustments, but also each jaw shaped in a steeply angled and opposed edge and exactly perpendicular to the axis of the collimator. The quartz prism also calls for much care and skill, for quartz is not only birefringent but also has the unusual ability to rotate the plane of polarized light in the direction of its optic axis, and such a prism must be made in two halves, one of right- and the other of left-hand rotating quartz, each cut so that the rays pass along the optic axis. These are then cemented together to form one composite 60° prism. Lenses must also be carefully computed and designed, again from quartz, and finally the plate holder needs to be very accurately made and angled to the spectral beam, since the smallest error in the positioning of the photographic plate will be detectable if the optical system is a good one.

Spectrographs are usually 'dog-leg' in shape, either in the horizontal plane (Hilger), with the plate-holder vertical, or they have the 'dog-leg' bend in the vertical plane, as in our Bellingham and Stanley spectrograph, Fig. 6.1, where it is concealed by the casing of the instrument, while the plate-holder and the slit are horizontal. Purchased for research on the composition of pearls, this instrument later proved invaluable for the detection of impurities in diamond powders, the composition of unknown minerals, transparency of materials to ultra-violet and for other purposes. Its dispersion is about 14 cm long, between the 'small' and 'medium' categories described above.

Fig. 6.1 The Bellingham and Stanley quartz spectrograph in the Gemmological Laboratory of the London Chamber of Commerce arranged ready for service with arc and quartz lens in position

The electrodes were pure carbon for alternating current, graphite for preference since they were freest from stray lines of their own. Or for direct current, metal (pure copper) electrodes which would allow checking for such elements as carbon which would obviously not be detectable with graphite rods. The numerous copper emission lines would serve as guides to approximate wavelengths of 'unknown' lines from the tested sample, an exposure for the copper rods being made before introducing the sample. The two spectra in close proximity on the same plate can then be compared very easily.

Using the main copper emission lines, a wavelength graph simi-

51

lar to the one illustrated in Fig. 5.2 was now prepared and the expected lines for the more usually encountered elements were added to a series of copies. With these as a visual check, the identities of such lines were simple enough to establish. As with visual spectra, the practised eye soon comes to recognize at a glance the presence on a spectrum plate of the pattern of lines belonging to the more familiar elements; no wavelength measurements being necessary.

But it is not possible even for the specialist to detect easily every element possibly present on the plate, especially when the element is an unexpected one showing perhaps two or three lines among many hundreds of others on the plate.

As an example, the spectrum of an unknown mineral was examined and the emission lines for calcium, silicon and a little lead were found, but we quite failed to notice the few sparse lines due to thorium which later was found to be a major constituent. Another laboratory, more experienced in mineral spectrum analysis, also failed to find this interesting and rare constituent, and a third laboratory, more used to dealing with glass than with minerals, only found it on the second attempt by very careful scrutiny of their spectrograms. The mineral in question had been more or less dismissed as an artificial glass for several years, but was now recognized as something completely new, the gem mineral ekanite, only the second metamict mineral to be discovered. It is now known to be fairly strongly radioactive, yet that fact did not emerge in the earlier tests.

Taking spectrum photographs

With regard to the technique of taking spectrum photographs, the same rules of strict alignment of arc, lens and slit with the optic axis of the instrument applied, as with the table spectrometer.

When seeking ultra-violet emission lines the lens was again of quartz. A spectrum of the pure electrodes only would always be photographed first, exposures with the arc about 40 cm from the slit being about five to ten seconds. The plate was then adjusted and the test material in powder form added to the lower electrode and the arc struck again. The 'thickness' or height of the spectrum was controlled by a V-shaped diaphragm: too narrow and it looked meagre and the lines possibly not well developed; too wide and the slight curvature of the lines due to the optical system became apparent.

A useful gadget was a metal screen slid into place over the

plate, with three rectangular apertures in echelon which could be screened to expose one at a time. One spectrum was taken through each in turn without need to change the position of the plate. This gave three narrow spectra in exact juxtaposition, ready for close comparison, which was not easily achieved if the plate had to be moved.

7 The Spectrophotometer

The principal changes in spectroscopy in the years since the original articles were written lie in the introduction of electronic recording instruments. In the past few decades the invention of transistors, semi-conductors, the silicon chip, lasers, micro-computers and other remarkable discoveries have made it possible to harness the optical spectroscope into a complex of ingenious electro-mechanical wizardry which, among other things, records the absorption on a graph.

This advanced instrument is known as a spectrophotometer but, despite the 'photo' in its name, it does not take pictures of the spectrum. It measures and records in graph form the amount of light transmitted at each wavelength. With all their undoubted sophistication, these machines are still essentially optical spectroscopes, although you would have some difficult in finding any way in which you could actually see the spectra which are being used.

There are many different models, some of them designed to perform specific tasks, such as astronomical analysis, forensic medicine, examination of incandescent matter, of infra-red or ultra-violet light, as well as the all-familiar rainbow of the visible spectrum. Basically all are similar in principle and it will suffice to describe in some detail the Pye Unicam PU8800 (Fig. 7.1), an instrument intended for general chemical analysis, which was purchased by subscription as a memorial to Mr Basil Anderson and is now on permanent loan from the Gemmological Association to the Gem Testing Laboratory of Great Britain, which is the latest name for the laboratory organized by Anderson seventy years ago.

We have so far concerned ourselves very largely with prism spectroscopes. In small hand instruments these give brighter and better-defined absorption spectra than those normally obtained through a diffraction-grating instrument. This is because the

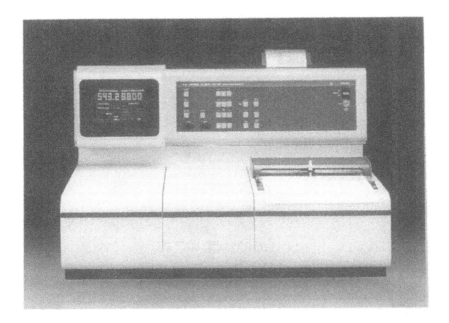

Fig. 7.1 The Pye PU8800

whole of the transmitted light is dispersed by the prism into a single spectrum, while in the diffraction instrument the light is split to give two first order spectra (one of which is used), two second order, two third order and so on until the reducing strength of each 'order' fades to greyness. Thus, even with today's advanced ability to extract the brightest possible spectrum from a grating the visual result must be less than half the strength of the incident light since only one first order spectrum is seen.

But in the spectrophotometer such overall faint light signals are vastly enhanced electronically and very accurately, and it is much more convenient to use the regular progression of wavelengths which are characteristic of diffraction-grating spectra, instead of the compressed-red and spread-blue wavelengths of the prism instrument.

In the PU8800 (see Fig. 7.2), there are two sources of light; a tungsten-halogen lamp for the visible wavelengths, and a deuterium arc for the ultra-violet region, the selected source being changed by a plane mirror which moves out of the optical path automatically at 325nm. The tungsten-halogen lamp has the property of redepositing vaporized tungsten on to its own filament,

55

and so avoids the gradual darkening of the glass envelope which would occur in a normal incandescent bulb. The deuterium arc takes the possible spectrum down to 190nm at which point atmospheric oxygen absorbs all shorter wavelengths.

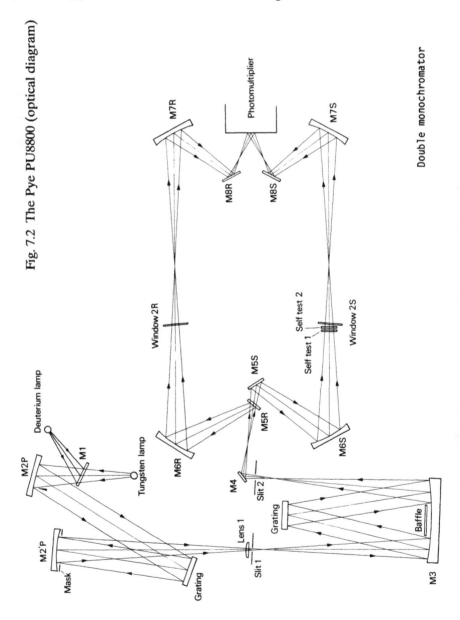

Fig. 7.2 The Pye PU8800 (optical diagram)

Light from these sources is modified by the automatic insertion of filters to reduce any contamination of test spectra by stray light, or by the intrusion of a second-order spectrum. A concave mirror focuses the light on to the monochromator entrance slit. (The Pye PU8800 used by the British Gem Testing Laboratory has an additional, fixed, diffraction grating between the light sources and the monochromator. This serves to increase the spread of the spectrum and enhances the sensitivity of the instrument.) Light then strikes one edge of a curved, or collimating, mirror which reflects it as a parallel beam on to the main diffraction grating whence it reflects back to the other edge of the concave mirror and on again to the exit slit. The two slits are selected from a number of different sizes cut in a disc in matching pairs so that the required width can be chosen by rotating the disc.

The diffraction grating is mounted on an arm connected with a drive spindle which gradually rotates through an arc, moving the spectrum it has produced slowly across the second slit and adjusting the focal distance at the same time, a complex movement which is synchronized with great accuracy with the paper feed in the recorder at the other end of the machine. Thus only a narrow beam of practically monochromatic wavelengths passes through this slit at any one time. I have said 'practically monochromatic' because with any slit size the light passed must consist of wavelengths within that slit width, so truly monochromatic light is not obtainable by this means. But at the narrowest slit widths this fact ceases to have any significance to the results obtained.

The monochromatic beam now passes through a beam-splitter, a pair of rapidly rotating, off-set, mirror segments which reflect it alternately into a reference beam, which irons out fluctuations of signal, and into a test beam which passes through, and is absorbed by, the specimen under test. The two beams are then re-combined, but not intermixed, by a further pair of rotating mirrors which are synchronized with the beam-splitter. They then pass through a powerful photo-multiplier cell which enhances the faint light signals enormously and converts the fluctuating beam of weak photons into a very intense but varying stream of electrons, in rapidly alternating bursts of reference strength and of a second strength reduced selectively by the absorption of the test specimen. These are passed to a detector cell which compares the component strengths and in its turn passes the resulting difference for each wavelength to the recorder.

The latter is a pen in contact with and travelling up or down as absorption changes, across a ruled paper graph moving laterally past it. The latter, by virtue of Unicam's special Synchroscan+ system, is fed through exactly in pace with the progressive but minute swing of the monochromator grating.

Throughout the instrument different sections are isolated in sealed casings and all windows between these are of silica glass to allow free passage of ultra-violet wavelengths. Mirrors of polished aluminium are also coated with silica to prevent oxidation and to allow U/V to pass.

In addition to the main recording apparatus there are various electronic recording features, including a display panel which monitors the progress of each scan and lists the particulars of the run. A print-out facility allows a permanent record of these to be made. Other details can also be recorded here, including a bar graph of the general absorption pattern. In the case of gem spectroscopy this may give no indication of the finer lines although it may serve to show which range of wavelengths need to be examined.

There are also facilities to allow a given section of the spectrum to be enlarged laterally, or for its absorbance record to be multiplied by factors from x2 to x10 in order to clarify small differences in absorption.

A single run on a stone may take up to twenty minutes, not counting the time employed in setting and masking the specimen suitably in the test cell, or time needed to interpret the resultant graph. So this is a far longer test procedure than the average gemmologist's quick squint down a hand spectroscope for an optical assessment of absorption. But it does provide a permanent and repeatable record which is far more accurate than the visual test and has the immense advantage of covering part of the near infrared, all the visible spectrum – without fall-off due to the fallibility of human vision – and the ultra-violet down to 190nm, and if an absorption line or band is there it will appear on the graph. There is no question of 'not being able to see the thing'. Nor can difficult lines be lost in a general murk at either end of the visible spectrum.

Glands in the light-proof outer case of the instrument allow feeds for additional equipment to be passed into the sample compartment, and in the Gem Testing Laboratory of Great Britain it has been found possible to take the tubes of the nitrogen gas apparatus through these to allow low temperature working (at about 120°K).

If needed, up to ten programmes may be stored in the PU8800, or complete records may be kept if a disc recorder is plugged into the instrument. But in most cases it is sufficient to store the paper graph together with a print-out of the information on the display panel.

In recent years physicists have been showing more interest in certain aspects of gemmology, although such interest does tend to turn mainly in the direction of fundamental science. For instance, diamond attracts their attention, not so much as a gem, but as a member of the small group of elements, including silicon and germanium, which are semi-conductors of great importance in micro-electronics.

These gentlemen are highly skilled scientists with access to university and museum laboratories which have spectrophotometers and other sophisticated equipment. The more authoritative gem-testing laboratories are also obtaining similar equipment, despite the costs involved, and published papers are tending more and more to represent spectra in graph form.

In the majority of cases the vertical (or ordinate) dimension in such a graph represents absorbance, and the horizontal (abscissa) dimension is in terms of wavelength. Some may have the abscissa marked or supplemented in EV values, which are favoured by physicists but are not so easily understood by the gemmologist.

Generally the graph will be arranged with the long wavelengths (red) to the right, matching spectra as published by the G.I.A., but in reverse to those seen in British publications – a minor difference which is easily reconciled.

We now have a graph on which the high, sharp peaks mean strong absorption lines, while lower 'bumps' will indicate broad bands in which the absorption is only partial. Both may be exaggerated electronically and a note may be included in the graph to say 'gain x 5', indicating that the vertical dimension has been multiplied by a factor of 5. This has no significance beyond emphasizing the absorption. The graph may cover wavelengths from inside the near infra-red – say 900nm – through the visible spectrum and down into the ultra-violet, but shorter ranges can be used. Electronic distortion known as 'noise' may reduce the record to a ragged peaked line at either end. In practice this can be ironed out reasonably well either by increasing the slit width or by speeding up the recording rate, or both.

Figures for 'band width' and 'path length' may be given. These are simply the slit width, usually quoted in nanometers, and the length of the light path through the tested stone, usually in millimetres.

A few graphs will be found in which the vertical dimension is in terms of transmittance, and in these the absorption lines will be seen as troughs rather than peaks. Again calling for an interpretation requiring little effort of mind.

To understand the graph correctly in terms of what may be seen in a hand spectroscope it is necessary to delineate the extremes of the visible spectrum, conventionally 700nm to 400nm. Some people may be able to see a little way past these two limits but that ability varies with individuals and acuity of vision is not good at the end wavelengths. Graph details beyond either limit are, of course, significant, often extremely so, for these parts of the spectrum tend to contain very relevant absorption lines which are not visible through the optical spectroscope. It is here that the instrument comes very much into its own.

The spectrophotometer is a complex machine and costs a great deal of money. Essentially it is a laboratory instrument and will be a long way outside the reach of the normal trade gemmologist. It has been discussed here in some detail for the sake of completeness and to aid the understanding of absorption graphs when they appear in gemmological literature. For most purposes the trade gemmologist will still depend very largely upon the ordinary small hand spectroscope, and that instrument is by no means superseded in the trade context by its more recent counterpart.

Part 2 Absorption of Gem Minerals

8　Absorption Spectra

Earlier chapters have attempted to outline the history of the spectroscope and give some account of its remarkable powers. The time has now come to embark on the detailed account of the absorption spectra of gem minerals which will probably be the most useful part of this book, and moreover will contain a good deal of original data representing the accumulated results of research over some twenty years in what was then known as the Precious Stone Laboratory of the London Chamber of Commerce, plus additional material culled from various sources since the original papers were published.

Some of the early work on the absorption spectra of minerals has been briefly mentioned in the historical survey, and credit will be given to the original discoverers of particular absorption phenomena (where these are known) when we come to deal with the individual species. Mention should be made at the outset, however, of the only previous worker who has attempted to publish a comprehensive survey of the absorption bands seen in coloured minerals. This was the American mineralogist, Edgar T. Wherry. Actually, he wrote two papers on the subject, each covering much the same ground. The later and more accessible paper appeared in the *American Mineralogist*, in 1929, in two parts, under the title, 'Mineral Determination by Absorption Spectra'.

The first part deals with minerals showing narrow absorption bands. This included zircon, but was mostly concerned with rare-earth minerals of no gemmological significance. The didymium yellow apatite, which shows so strong a spectrum, was not noted by Wherry. The second part covered minerals showing broader bands, and these are grouped in what is probably the easiest manner for reference – according to colour. Wavelengths are given in millimicrons (1 millimicron = 1 nanometer) and Wherry's own measurements are given in each case, together with figures derived from other workers where these had been published. The instrument used by Wherry was an Abbé-Zeiss 'spectroscopic eyepiece', with wavelength scale attached, and in his paper there are valuable suggestions as to the technique which should be employed to get the best results from such an instrument.

Despite Wherry's pioneer efforts to popularize the spectroscope as a rapid means of identifying many mineral species, his advocacy had apparently very little effect.

Gemstones, far more than minerals, as a whole, lend themselves admirably to this method, since most of them are both coloured and transparent, and a high proportion of them show recognizable absorption bands. Nevertheless, when we first started our experiments there was no hint in any textbook that this was the case. Only in a few books were there accounts, always inadequate, of the bands first observed by Church in almandine and in zircon. Only in the case of these two minerals, one was led to assume, could the spectroscope be usefully employed by the gemmologist.

Daily 'Discoveries'

A large green zircon, having a strikingly strong spectrum, was indeed the stone in which we first observed with pleasure and excitement absorption band phenomena. But when, having worked out our own technique for observing the bands in this easy spectrum to best advantage, we turned our instrument on to ruby, emerald, spinel and other coloured gems, we were astonished to find how many of them displayed bands which were at once distinct and distinctive. Only after this stage had been reached, and daily 'discoveries' made, did we begin to search the literature and become aware, not only of Wherry's work, but of excellent researches on individual minerals carried out by other scientists and completely neglected or forgotten.

But from the gemmologist's point of view a whole new world lay waiting to be explored, and during the next few years we were exceedingly busy measuring and recording absorption bands in all the gem minerals we could lay our hands upon. We found the new method not only scientifically interesting and aesthetically satisfying, but of enormous practical value in the routine testing of gemstones. Filled with zeal, we began to teach and preach the new method. The seed was sown amongst students at Chelsea Polytechnic and a number of these mastered the technique, though others found it difficult and did not persevere. That good friend of the student, Webster's *Compendium*, made its appearance in 1938, and contained a useful section on the more important absorption spectra, illustrated with diagrams. Our own resumé of the subject was handed to Dr Herbert Smith at about that time for inclusion in the rewritten ninth edition of his classic *Gemstones*, which finally appeared in 1940. In this we gave an adequate account of all those spectra which have first-class diagnostic importance. Only in the case of zircon did we attempt a full-scale treatment on the lines now offered. The subject had grown under our hands to formidable proportions and there seemed no finality. Time was increasingly taken up with routine work – and then came the war.

In the 1950s we decided to attempt a full account of all the spectra we had observed.

Mere catalogues of wavelengths make dull reading, and some form of illustration is most desirable. This presents considerable difficulties. One would expect that colour photographs would provide the best answer, but the inherent absorption factors and relative imbalance of colour film emulsion to different wavelengths tend to give a false rendering of the visible spectrum and are not suitable.

Coloured drawings which appeared in the second edition of the *Compendium*, and as the frontispiece of Anderson's *Gem Testing*, were drawn to represent the bands as seen through a diffraction-grating spectroscope, whereas those who used the method most almost invariably used a direct-vision prism instrument, which alters the apparent distribution of the dark bands. In this book we will illustrate each important spectrum three times in black and white, showing the dispersion, first as seen through a prism instrument, then through a diffraction spectroscope, while the third version will give the reversed prism spectrum (red to the right) which is favoured by American gemmologists. (The vertical width seen through different makes of instrument may vary, and in general will be greater than that illustrated, so that lines and bands may seem closer together than in the illustrations.) With a little imagination the monochrome drawings will be understood, especially if the colour guide under each is considered; the wavelength scale at the top edge should also help. The colour divisions are as seen by the normal human eye, and each colour changes gradually to the next as we progress through the spectrum. There is, for instance, no exact wavelength at which yellow ends and green begins, and basic colours can be subdivided even further, but the boundaries are still arbitrary and opinions may vary as to their exact position. But the colours indicated should give guidance to the positions of absorption lines or bands when seen against the full rainbow of the spectrum. The intensity, breadth and 'pattern' of the lines and bands are the important things to memorize and recognize.

Before proceeding to a description of the actual spectra, a few words must be devoted to the necessary technique and the types of instrument recommended. Both are capable of considerable variation according to the choice and circumstances of the operator, but a brief description of the procedures which we have found most satisfactory may help the beginner. There are three essentials – a powerful light-source, some convenient means of transmitting the light through the specimen (or reflecting it from the specimen) so that it passes into the spectroscope slit, and finally the spectroscope itself. It has been found in practice that for general observational purposes a small hand spectroscope, either of the prism or grating type, gives the most satisfactory results. Built-in 'microspectroscopes' have the advantage of stability, and may seem easier for the beginner to use, but they are more expensive and actually less flexible in use than a spectroscope held in the hand.

Prism instruments give clearer and brighter spectra, but must be carefully chosen, since if the dispersion is unduly short the bands in the red region will be too cramped for details to be seen; while a very wide dispersion is also to be avoided if the bands in the violet are not to become so spread and diffuse that they are imperceptible. If a Victorian brass-cased prism spectroscope is selected, it is important to check that it does cover the whole visible spectrum. One or two have been known to cut-off short at the blue end. Grating spectroscopes have the advantage of even dispersion, but usually do not give the same intensity of light against which to see the absorption bands.

The original articles confidently recommended several Beck instruments which unfortunately are now out of production. But the construction diagrams of two of these have been retained here (Figs. 8.1 and 8.2) since they serve to illustrate the general structures of any similar spectroscopes.

If a secondhand Beck of either pattern should come on the market in good condition, then it is worth considering.

For wavelength measurements a Beck wavelength grating spectroscope was used, but this again is out of production today, as is the Hartridge reversion spectroscope which gave two images of a spectrum, the one reversed against the other, so that a line or band in one spectrum need only be brought into coincidence with its counterpart in the other to allow the relevant wavelength to be read from a vernier on the drum which moved the two spectra in relation to each other.

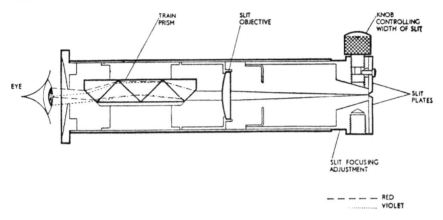

Fig 8.1 Construction of Beck No. 2458 prism spectroscope

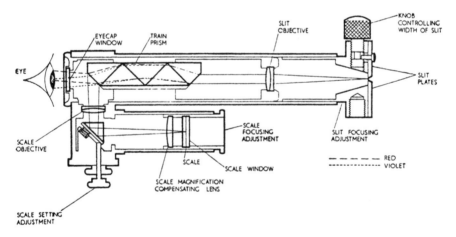

Fig. 8.2 Construction of Beck prism spectroscope with wavelength scale

Using such instruments enabled the two workers to arrive at wavelength readings of considerable accuracy. But where the intensity and sharpness of lines allowed they confirmed these values by using a table spectrometer together with a diffraction grating and comparing the bands with emission lines of known wavelengths in the vicinity of the band to be measured.

Available spectroscopes

Today the Gemmological Association can supply three different versions of the Meiji Techno prism spectroscope, each of which is suitable for most gemmological work:

- SP 100. A simple prism instrument with sliding focus and fixed slit.
- SP 150. As the SP 100 but with an adjustable slit.
- SP 200.05. As SP 150 but with a wavelength scale.

They also carry a stock of the tiny, comparatively low priced, OPL diffraction-grating spectroscope, which provides a useful back-up to the prism instrument in that the blue wavelengths are less spread and perhaps easier to see. In this connection my attention has been drawn to the basic fact that as one becomes old there is a slow yellowing of vitreous humour of the eye and it is less easy to see the shorter blue and violet wavelengths. For this reason older gemmologists may have trouble seeing bands caused by ferric iron, or some due to manganese. This may at least in part be overcome by increasing the illumination.

A minor disadvantage of the rather costly wavelength type of spectroscope is that some provision has to be made to illuminate the scale separately from the stone under test. Also the scale is not always accurate for all parts of the spectrum, especially when the focus is altered.

Dr Gübelin had a spectroscope designed and constructed specially to serve the needs of the gemmologist. This has built-in lighting, an illuminated wavescale, spring tongs for holding the specimen and other refinements.

The Gemological Institute of America also markets a special assembly which, in its most recent form, uses an optic fibre light source with interchangeable red, green and blue filters to isolate those sections of the spectrum for easier examination. There are other similar assemblies available but the principal disadvantages of them all are their high cost, and a certain lack of flexibility for examining mounted specimens, opaque stones by reflected light, or under different types of illumination, for which the simple hand-held spectroscope is so well adapted.

Whatever the instrument, an adjustable slit is virtually essential. When there is plenty of light the slit should be as narrow as possible – just open enough to eliminate horizontal streaks caused by dust or minor irregularities on the slit edges. Most modern instruments have a glass cover which helps to minimize this problem. The slit may need to be opened to admit more light when dealing with strongly absorbing stones, but this should not be overdone as the wider the slit the more diffuse the bands become owing to the overlapping of an infinite number of slit images along the spectrum.

In general the narrower the slit the better and if light intensity can be increased, this is preferable to opening the slit.

Focusing is the other important adjustment. Test this by looking through the instrument at the Fraunhofer lines in daylight, adjusting the slit width and the focusing slide until the main dark lines are seen clearly and distinctly. Alternatively the emission lines from any fluorescent lighting tube can be used for the same purpose.

9   Light Sources

We now come to the question of suitable forms of illumination for the absorption spectroscopy of gemstones. For satisfactory results with specimens of all degrees of transparency and colour a really powerful source is necessary. Further, arrangements should be made to examine the spectrum not only of light transmitted through the specimen, but also of light internally reflected or scattered from it.

Direct transmitted light is usually the best for stones of fair transparency having strong absorption bands. The reflected light method gives stronger absorption effects, and is well-suited to pale stones with feeble spectra; scattered light is the best where fluorescence spectra are concerned.

In practice it was found that these requirements were best met by a 500-watt projection lamp, and this was housed in an asbestos-lined box with a 10 cm square hole in one side. Today, with asbestos regarded as a serious health risk, a simple slide projector would probably fill this role just as efficiently, especially if it were fan-cooled. These high-wattage lamps generate considerable heat, so burns, and even fire need to be guarded against. If light and heat are further concentrated on to a stone – by passing through a substage condenser, for instance – damage and colour changes are possible in some gemstones.

In unfavourable conditions it is usually possible to get enough light into a stone by using a 100-watt clear bulb at very close range, say 10 cm, again guarding against burns, and perhaps concentrating the light even more by the judicious use of a simple lens. But such lighting is unlikely to be enough for dark, highly absorbent stones, when the extra power of a 300-watt or 500-watt source is necessary.

Fibre optics provide another means of powerful illumination today, and they have the extra facility of being quite flexible, allowing strong lighting to be brought right up to the stone tested. But this type of lamp is expensive, and although it conveys visible light very well, glass or plastic fibres will not transmit UV wavelengths anywhere near so efficiently and such a source is not so good when fluorescent lines are looked for.

The next problem is to concentrate a beam of this light through the specimen and to project it evenly on to the slit of the spectroscope. One *can* get results of a kind simply by holding the stone in tongs in front of the slit in the manner of the dichroscope, but the effects are so patchy and streaky that detailed observations are impossible. The most satisfactory general method is that first suggested by James Payne, making use of the microscope with its body tube vertical, fitted with a low powered objective (say 35mm) and the eyepiece removed. The stone to be tested is placed on a glass slide on the table of the microscope, light from the lamp is concentrated on to the mirror, and this, the sub-stage condenser, the stone itself, and the focus of the microscope are all adjusted until the brightest possible even glare is passing directly up the tube. The spectroscope, held in the hand, is then rested lightly on the microscope tube in place of the eyepiece and an excellent streak-free spectrum should be seen.

There are several notes to be made here. For concentrating light on to the microscope mirror from a projection lamp we strongly recommend a flat-bottomed round flask filled with water – 600 cc is about the most suitable capacity. This is simple, stable, cheap, has great light-gathering power, and above all, filters out the unwanted heat rays which may otherwise cause damage to stones on which rays from the lamp are concentrated, or, at any rate, will make them uncomfortably hot to hold.

Another reason for keeping the stone cool is that all absorption bands tend to become weaker and more diffuse when the temperature is raised.

At this point the original paper recommended that a second flask containing a strong filtered solution of copper sulphate in distilled water should be kept to facilitate seeing the blue and violet end of the spectrum and for observing fluorescent effects in the red. Copper sulphate in our modern world is less easy to obtain, and apparently one can no longer simply order it from the high street chemist.

Alternative 'fluorescence exciter' glass filters are now available and should be used when a fluorescence effect is sought in the red, or when it is desirable to screen out the red and orange to enable the shorter wavelengths to be seen.

Another useful hint in adjusting the light passing through the specimen to best advantage is to place a piece of ground glass on top of the microscope tube. This enables one to find the optimum position without dazzling the eye with the strong glare coming up the tube. Alternatively, glare during adjustment can be diminished by viewing through a dark colour filter such as the Chelsea filter. With stones of moderate refractive index, enough light can usually be transmitted when they are simply placed table-facet down on the microscope. With highly refractive stones such as diamond, there may be too much total reflection to allow enough light through. In such cases it will be better to allow the stone to rest on a pavilion facet, which will enable light to pass through one of the crown facets nearly parallel to itself. In really difficult cases the specimen may be immersed in a suitable liquid. With small stones diaphragms of metal or other opaque material are useful to ensure that all the light viewed has actually passed through the stone; white light which passes around its edges will 'drown' the absorption bands if this point is not attended to. Larger stones usually need no diaphragm.

The inside of the microscope tube should be matt black. If this is not so then it should be lined with black flock paper to remove possible extraneous light reflections.

The reader may think that all these arrangements and adjustments sound fussy and elaborate, but they are given in some detail as they all contribute to success in using the spectroscope to good advantage. Once the lamp, condensing flask and microscope have been placed in position it is a matter of seconds in practised hands to place the stone on the microscope, tilt the mirror to best advantage and view the spectrum. For diagnostic purposes with the average run of stones extreme niceties of adjustment are seldom necessary.

When examining a stone by reflected light, the microscope is lifted to one side, the stone to be tested placed table facet down on a piece of black cloth in the focus of the condensing flask, and the light reflected from inside, usually from the back of the table facet, is viewed through the spectroscope by holding it at an angle of about 45° with the slit an inch or two from the stone. When seeking for fluorescence lines (in ruby or spinel, for instance) the best effects are seen when the stone is viewed from the side (scattered light). Whether the stone is on the microscope or on the bench does not greatly matter for this.

In the case of specimens which are too opaque to transmit enough light for observation of absorption bands one can sometimes detect these in light reflected from the surface of the stone. This is a useful technique in dealing with jade ornaments and with turquoise.

This leads to an alternative to the microscope set-up, using a simple stand to hold the spectroscope at an inclination of 45° with the slit about 6 cm from a small matt-black covered rotatable stand on which the stone is placed, table facet down. Light, with the possible addition of a bull's-eye lens to condense it to a brilliant spot, shines at a similar angle but from the other side of the stone, so that it is totally reflected from the back of the table facet and up into the spectroscope slit. By placing the latter a short distance from the stone the transverse lines resulting from the faceting are largely eliminated. This method allows surplus white light to be absorbed by the black-covered stand, which is turned to give the best possible spectrum, while light reflecting into the instrument does so more directly and on a longer path through the stone than in the microscope method. The absorption spectra therefore tend to be better defined.

Another method of illumination was suggested by the late D.J. Ewing of Edinburgh. He enclosed a car headlamp in a large tin. A condensing lens above this concentrated the light on to a small hole in the centre of the lid and the stone was placed over the hole. The spectroscope was held about 2.5 cms above the stone. Unfortunately the lamp cover tended to get very hot and needed fan-cooling of some kind. It was simple to make and gave brilliant spectra with the advantage that the instrument could be tilted at different angles. Again it could be used with the spectroscope further away, which eliminated transverse lines. Contrary to what might be expected, this removal of the instrument by five or seven cms made little difference to the amount of light entering the slit in either the Ewing instrument or with the stand previously described.

When invigilating practical examinations one has occasionally been amazed to see an examinee dash in, seize a spectroscope with both hands and immediately twist the slit adjustment knob while squinting through the instrument at some inoffensive gem. Too often the net result was that the candidate got a magnificent display of bright spectral colours, because he had opened the slit as far as it would go, but could see none of the finer absorption lines because they were masked by an infinite number of wide slit images along the whole observed spectrum. Invigilators are not allowed to point out errors, and the class instructor should have corrected this much earlier.

The advised procedure in any circumstance where more than one person is using a spectroscope is to assume that the last person to use the thing has left it incorrectly adjusted. Check slit width and focus by looking through it at a sodium lamp, or at any fluorescent light tube, or even at the Fraunhofer lines in daylight, sliding the spectroscope tube to a correct focus and making the slit as narrow as possible without cutting off the spectrum completely. Once these two adjustments are achieved, resist all temptation to alter them, unless the dark colour of a stone forces you to admit more light. Even then it is far better to increase the light source or to move the stone to allow more light to pass through a thin edge.

As with other instruments it is a good idea to have a little flexibility of movement. A stone which at first yields nothing in the way of absorption, may do better if it is turned round or turned over. The spectroscope perched on top of the microscope or in its inclined stand may need a slight tilt one way or the other to get the best beam of light for our purpose.

In the case of very faint absorption lines, where one is not sure whether imagination is taking over, it pays to line up the instrument, stone and light source and then to move the stone to one side so that one is looking at the complete and unabsorbed spectrum. Now push the gem back into the light path while still watching the spectrum. Any faint and narrow absorption will then become visible as the stone moves into place, and can be checked again and again at will, by moving the stone in and out of the light beam. The eye is always more sensitive to movement and this method achieves this in the observed spectrum.

In either type of instrument the focus changes gradually with the wavelength, so that the draw tube should be pushed in a little as the blue end of the spectrum is approached. The difference in the case of small hand instruments is quite small and focus may not need to be adjusted unless very accurate measured readings are being made with a measuring spectroscope. The usual practice is to get the focus right for the red end of the spectrum, where fine lines such as those of chromium may be expected; bands in the blue are generally broad and need not be quite so carefully focused.

It has already been said that, where a stone shows only feeble absorption bands, clearer results may be obtained by examining light reflected from inside the stone, as the path through the stone is then longer than when light passes straight through. But even in the latter case absorption can often be increased in strength by turning the stone on its side, or even on end. If the stone is a large one it can be held in the fingers; but one must always guard against confusing the spectrum of haemoglobin, the red cells of the blood, with that of the stone. A far better way is to use a small piece of the plastic adhesive, Blu-Tack, to hold the stone in the required position. This stuff can be very useful indeed in gemmology. For instance if a diaphragm is needed to mask out extraneous light, make a hole in a flat piece of Blu-Tack and press the stone firmly in contact over the hole.

Quite apart from variations in the strength of absorption due to the length of the transmission path there are of course important differences due to pleochroism. A strongly pleochroic stone such as alexandrite really has three absorption spectra, though these can be seen separately only by using polaroid or a nicol prism. Such effects due to dichroism seldom interfere with the recognition of the total absorption pattern of the gem. On the other hand, a weak absorption in a birefringent gem may be made to appear stronger by turning a piece of polaroid to its optimum position over the eyepiece of the instrument.

10   The Causes of Colour

The causes of colour in minerals are many, and often are not properly understood. Under certain conditions elements not usually regarded as one of the 'colouring agents' of inorganic nature can give rise to hues of great intensity – as for instance sulphur in the mineral form of ultramarine known as lapis lazuli – while on the other hand certain compounds of iron and copper (metals well-known as colouring agents) are quite devoid of colour.

The number of elements known to produce definite absorption bands in the visible spectrum is very limited, and most of the bands which we see in gemstones can be ascribed with some confidence to one or other of these elements. Most important are the so-called 'transition elements', titanium, vanadium, chromium, manganese, iron, cobalt; nickel and copper, which occupy adjacent positions in the periodic table of the elements in the order given, from atomic number 22 (Ti) to atomic number 29 (Cu). A few of the rare-earth elements also give rise to well-defined bands and so, under certain conditions, do nitrogen (N) and uranium (U).

Fortunately for gemmologists, the position of the absorption bands due to a given element varies considerably according to the species of the host mineral. Thus, when examining the spectrum of light transmitted through ruby we are able to say not merely 'chromium is present' but 'this is the spectrum of chromium in corundum' – or in other words 'this is a ruby'.

Only in idiochromatic minerals (where the colour is due to an element which is an essential part of the composition and is therefore constant within narrow limits), such as almandine and peridot, can the absorption spectrum be said to characterize the entire species. In allochromatic minerals, which form the majority of gemstones, the absorption spectrum will differ with each colour variety, since different elements are usually causing the colour in each instance. Thus, while red corundum and red spinel display

76

bands typical of chromium, the blue varieties of these minerals show completely different sets of bands, due to iron. Beryl and chrysoberyl may be cited as further examples of minerals which can show either a chromium or an iron spectrum.

It is worth remembering that, almost invariably, a chromium spectrum betokens a mineral in which aluminium is an essential element, since it is by small-scale isomorphous replacement of Al_2O_3 that Cr_2O_3 usually enters into the crystal lattice of gemstones and gives rise to colour. In a similar way, when the unmistakable didymium spectrum is seen it can be assumed with confidence that the mineral contains calcium, for the two rare-earth elements collectively known as didymium are so chemically akin to calcium that they follow it with dog-like devotion throughout the natural processes of reaction and crystallization in which minerals are formed. More will be said on this topic in the appropriate place.

The time has now come when we can enter upon descriptions of the absorption spectra of all those gem varieties in which we have been able to observe bands. The order in which these should be treated is not an easy matter to decide. For convenience in practical testing a grouping according to colour is undoubtedly the most effective – that is, to describe the different spectra seen in all red stones, all green stones, etc. This grouping is convenient for live classes and was that adopted by the Post-Diploma class at Chelsea Polytechnic. But initial teaching was based on gems grouped according to the known colouring element; spectra due to chromium, spectra due to iron, and so on. In our chapter in Herbert Smith's *Gemstones* (1940) we adopted the latter as the more scientific grouping.

Unfortunately the causes of absorption bands in some gems are not known with certainty and one either has to commit the scientific indiscretion of over-confident guessing or allow for a 'miscellaneous' group where such mysterious spectra can all be included. Yet another possible arrangement is to use alphabetical order. This is followed in Chudoba and Gübelin's *Taschenbuch* (1953) and, though entirely arbitrary, is very convenient for reference when one knows the name of the stone and wants details of the spectrum.

In this book all of the methods described will be employed in their turn. First, in order to ensure a rational sequence in which similar spectra will be grouped together, they will be described according to the elements which cause the absorption bands. These descriptions will be given in much greater detail than has been previously attempted, and will include many bands not hitherto recorded.

When these detailed accounts have been completed a summary of the spectra will be given, grouped according to colour, and giving only those bands which are diagnostically important. A list in alphabetical order will also be given.

Among the eight transition elements listed earlier, neither titanium nor nickel seem to produce definite absorption bands, though they undoubtedly have an influence on the colour of compounds in which they appear. Of the remainder, chromium and iron are by far the most important; one or the other, and often both together, play a major role in the colours seen in the world of gems, and each gives rise to distinctive absorption bands. We propose to deal first with the spectra of gem minerals coloured by chromium.

Chromium spectra

Chromium is the aristocrat of colouring metals, and produces the finest reds (ruby, spinel) and the finest greens (emerald, jadeite) amongst the gem minerals. Iron, too, can produce quite handsome reds and greens, but colours due to chromium have a brilliance and clarity which those due to iron cannot match. This can be explained on the grounds that the absorption regions in chrome-coloured stones are sharply defined, while the colours outside these regions suffer hardly any absorption, leaving the hues at full strength. In the case of iron, the possibilities of electronic energy-changes exist to some extent in all parts of the spectrum, so that even the colours transmitted have suffered considerable absorption, resulting in a general dimming or greying of the resultant colour. The chromium absorption bands are like towns which have no suburbs, so that the open country between them is left fresh and undefiled – whereas with iron the main towns (bands) have no clear limits and a certain suburban haze is over all.

Recognizing a chromium spectrum

Absorption spectra due to chromium have been studied more thoroughly than those due to any other element, with the possible exception of some of the rare-earth metals. Despite the very different hues produced – red, green, purple – the spectra have a great family likeness and the practised eye can recognize a chromium spectrum at a glance, even in an unknown mineral. In addition to broad absorption regions in the green and violet, the position, strength, and extent of which determine the colour, many of the bands due to chromium are of extreme fineness meriting the description 'hair-lines' which has been applied to them. The production of such unusually narrow absorption lines is apparently due to the fact that the excited electron is deep down in the chromium ion and is thus relatively unaffected by the electric field of the surrounding ions. A group of these fine lines in the red end of the spectrum is indeed the characteristic feature of a chromium spectrum and, though their position varies with each mineral, there is a great similarity in their grouping. The most prominent of the lines is always a close doublet (usually too close to be resolved in a hand spectroscope) and there are usually two moderately strong additional lines towards the orange side of the main doublet. In several species a number of fainter lines can be observed in the red, and further narrow lines in the blue region are also a feature of most chromium spectra.

Another distinctive and remarkable phenomenon connected with chromium spectra is the 'reversibility' of the narrow lines in the red, which can be seen as bright emission lines due to fluorescence or as dark absorption lines, according to the conditions. A brightly lit ruby, for instance, is absorbing and emitting light simultaneously at the same wavelength in a manner reminiscent of glowing sodium vapour. The presence of iron, that great killer of fluorescence, is apt to mask this extraordinary effect in some cases and even an excess of chromium can have a similar action. Fluorescence lines can be far more sensitive than absorption lines as a test for the presence of minute traces of chromium, especially in corundum and since they have a great diagnostic value and are inevitably seen when studying the absorption spectra of certain gemstones, we shall treat with these as they occur.

Gem species or varieties in which we have observed absorption bands due to chromium include the following:

Red or pink stones: Ruby, spinel, pyrope and topaz.

Green stones: Alexandrite, emerald, jadeite, nephrite, demantoid, uvarovite, diopside, enstatite, euclase, kyanite, hiddenite, aventurine quartz, stained chalcedony and glass. In some of these the more important bands are due to another element, and will be described again under that element.

11 Absorption and Fluorescence Spectrum of Ruby

Observations on the luminescence spectrum of ruby preceded by many years any account of the closely related absorption lines. As long ago as 1859 Edmond Becquerel[1] described the bright lines in the red end of the spectrum emitted by calcined alumina, and by natural corundums in different degree, when placed in a shaft of sunlight in his ingenious phosphoroscope. Since every specimen of corundum and of calcined alumina that he tested showed the same spectrum, varying only in intensity, Becquerel assumed that these luminescence lines must be a characteristic feature of alumina itself. He did, however, notice that precipitated alumina did not give the red glow until it had been strongly heated for a prolonged period, and that when some chromium salts were added to the alumina prior to this treatment the red glow was much enhanced in the resultant specimen.

For nearly thirty years an argument was carried on in the form of scientific papers holding opposite views as to whether the presence of chromium was essential for the red luminescence of alumina to appear. So redoubtable a scientist as Crookes,[2] for instance was of the contrary opinion. To Lecoq de Boisbaudron[3] goes the final credit of proving that chromium was indeed the activator in this case: he was able to show that as little as 1 part of Cr_2O_3 in 10,000 Al_2O_3 was sufficient to give rise to a perceptible red fluorescence.

The *absorption* bands in ruby were apparently first described by A. Miethe in 1907.[4] He observed and measured the red doublet, the two weaker lines on the short-wave side of this, and the three strong narrow lines in the blue. In 1908, Moir, working independently in South Africa, made some careful measurements of the ruby spectrum, but did not understand the nature of the main fluorescence doublet, which he mistook for a narrow region of intense transmission traversed by a dark 'hair line' due to absorption. Keeley, in 1911,[5] also thought that the fluorescence line was a transmission region between two absorption bands, but in 1915 interpreted it correctly. Moir, incidentally, was so greatly impressed by the beauty and diagnostic possibilities of the ruby spectrum which he studied soon afterwards, that he searched specimens of sapphire, rubellite, red spinel, aquamarine and topaz for absorption lines, but in vain. Probably the instrument he was using had so large a dispersion that only the strongest and most clearly defined bands could be seen: a fact which emphasizes the importance of suitable instruments if this form of gemstone identification is to be a success.

Amongst others who worked on the fluorescence spectrum of ruby were Dubois and Elias[6] (1911), Mendenhall and Wood[7] (1915) and Deutschbein[8] (1931). Deutschbein carried out an elaborate series of measurements both at room and at liquid air temperatures on ruby and many other chromium phosphors.

When, early in 1933, we began our own observations on the spectrum of ruby we knew nothing at all of any previous work on the subject, and had all the excitement and surprise of seeing the bright doublet in the red and proving for ourselves that it was due to fluorescence.When we first saw it, we thought that it was unique to the particular specimen we were testing! Only later did we slowly become aware of how much work had already been done, a small part of which has been mentioned above. Nevertheless, in our subsequent researches we have at least had the satisfaction of recording some bands which have not been previously noticed by other observers.

Here, then, is the full tally of bands and lines in ruby that we have been able to measure. Not all of them have been seen in any one spectrum.

We use seven grades of strength in all, though these are purely relative, since in pale or small specimens even a strong line may be weak in appearance. So in the list s indicates 'strong', m 'moderate', w 'weak', while v adds the qualification 'very'.

F signifies a line which in the right conditions of lighting may be seen as brightly fluorescent, while f lines are those less easily seen as fluorescent, and may need the red and orange wavelengths of the light incident on the stone to be screened out by a copper sulphate filter to make them visible. Otherwise they may appear as dark lines under unfiltered lighting conditions. A fluorescent line may sometimes be seen as a dark one, or vice-versa, simply by shifting the angle of the spectroscope a little.

f	712nm	vague	w.
f	705.5nm	fine	m.w.
f	702nm	''	m.s.
f	699nm	''	v.w.
f	697nm	''	v.w.
F	**694.2nm**	''	v.s.
F	**692.8nm**	''	v.s.
f	688.5nm	''	m.w.
f	673nm	''	m.w.
f	**668nm**	''	m.
f	**659.2nm**	''	s.
	646nm	''	v.w.
	641nm	''	w.
	626nm	''	v.w.
	610–500nm	broad	s.
	481nm	fine	v.w.
	476.5nm	''	v.s.
	475nm	''	s.
	468.5nm	''	s.
	463.5nm	''	m.w.

This list may appear formidable but it records all the absorption and emission lines we have been able to observe in ruby. In specimens of normal size and colour only the seven in heavy type, and the broad absorption in the yellow and green, and the strong absorption of the violet, are commonly observable, and these are ample to identify a stone as ruby, but of course they will not separate natural from synthetic.

In the first and third of the three diagrams (Fig 11.1) of these basic lines it will be noticed that although the distance between the lines in the blue doublet and in the red doublet is practically the same in terms of wavelength, the red doublet can be seen as only a single line whereas the blue lines are clearly separated due to the greater dispersion at this end of the spectrum. This difference is removed in the diffraction version and both doublets may be observed as such if great care is used.

Ruby being a strongly dichroic stone, it is natural that there should be distinct differences between the absorption of the ordinary and the extraordinary rays. If we rotate a polaroid disc in the light path while observing the spectrum it will be seen that the central broad absorption is relatively insignificant in the extra-ordinary ray, allowing some yellow to pass which spoils the colour of the stone in that direction. Then as the polaroid disc is turned through 90° the shadowy band intensifies and widens by nearly 20nm towards the orange, until the full massive absorption is seen in the ordinary ray position. If during the same process, the attention is fixed on the blue doublet, an interesting variation in the intensity of these two lines can be noticed: 476.5 is stronger in the extraordinary ray, 475nm in the ordinary; also the rather vague line at 668nm is missing in the extraordinary ray, while 659.2nm is stronger. The red doublet, whether absorbing or fluorescing, is stronger in the 'ordinary' spectrum.

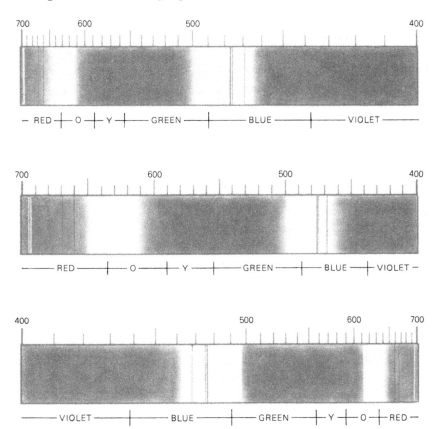

Fig. 11.1 RUBY. Both natural and synthetic rubies have the absorption printed in heavy type in the text. The bright emission doublet at 693.5nm may be seen as a dark absorption in some circumstances

Since the ordinary ray must invariably be present in all orientations of ruby it is this spectrum which predominates under normal conditions of observation, and the variations mentioned above, though interesting to the physicist, are of no diagnostic importance.

It has been stated by Wherry and others that synthetic rubies can be distinguished from natural stones by the greater intensity and width of their absorption bands. It is true that synthetics usually carry rather more chromic oxide than stones from Burma, Siam, or Ceylon and therefore give a more striking spectrum; but though we are keen advocates of the spectroscope we would not recommend any reliance on this method of distinguishing natural from synthetic rubies. There are chrome-rich stones from Burma, for instance, as well as rubies from the Tanzanian source, which have as rich a spectrum as any synthetic.

In deep-coloured stones the absorption region in the violet extends almost up to the 468.5nm line, and spectrum photographs have shown us that this powerful band extends well into the near ultra-violet. There is then a further region of transmission before the ultimate absorption sets in, which in natural rubies is at between 300 and 290nm, but in synthetic stones not before 270nm as a rule.[9]

Since the fluorescence lines are a more sensitive and diagnostic test for ruby (or for any corundum containing a trace of chromium), it is well to develop the quite simple techniques needed to see these effects at their best. Scattered light shows up the fluorescence doublet more clearly than direct transmitted light, chiefly because the continuous background is not so bright, and there is thus greater contrast.

Still better results can be obtained by using a copper sulphate filter which will allow the deep red fluorescent lines to be seen against a totally dark background, since copper sulphate transmits light only up to about 550nm and therefore absorbs all red light from that which falls on the ruby. A nearly saturated solution of 200 grams of cupric sulphate in 600 millilitres of distilled water will give a blue liquid suitable for this and for other absorption tests.

The solution should be filtered to ensure clarity, and poured into a 600 cc flat-bottomed spherical chemistry flask, which then serves as the condensing lens at the same time as providing a most efficient blue filter.

Since in recent years it has become rather less easy to obtain copper sulphate through the ordinary high street chemist, a glass filter has been perfected by McCrone Research Associates Ltd of Hampstead, which is almost the equivalent of the $CuSO_4$ one, but cuts off light a little nearer the red, at about 600nm. This and a high barrier red filter are obtainable, but both are small in size, intended for use with a pen-type light source which is marketed by the same firm. The blue one on its own, or together with the deep red, is as efficient as the Cu sulphate solution.

As will be described in the next chapter, recognition of red spinel and its distinction from ruby by spectroscope depends very much on the arrangement of their fluorescent lines, and this task is made much easier if a blue filter, of either type, is used.

Summing-up

The absorption spectrum of ruby is a typical chromium spectrum, and is highly diagnostic. Its most constant features are a strong doublet in the deep red, appearing as a single line in a small prism spectroscope, and usually seen as a *bright* fluorescence line: two less prominent lines to the orange side of this: a broad absorption region covering the yellow and green: a 'window' in the blue in which a powerful absorption doublet and a third strong line can be seen, followed by general absorption to the end of the spectrum. In experienced hands, the fluorescence doublet alone is proof of ruby, but a glance into the blue to confirm the presence of the three dark lines in characteristic grouping forms a valuable safeguard against possible error. While some of the other chromium-coloured gemstones have absorption spectra which bear a certain resemblance to that of ruby there should be no real danger of confusion. The absorption bands of alexandrite are the most similar, but the strong doublet in the red is only feebly seen as luminescent, and the narrow lines in the blue have not the same grouping. In any case, the appearance of the stone should act as a guide in this case. Spinel has a different structure of fluorescent lines, and lacks the lines in the blue, and while fine pink topaz may show a fluorescence line, this is exceedingly feeble, is not accompanied by the accessory lines in the red as in ruby, and shows no lines in the blue.

12   Absorption and Fluorescence Spectrum of Red Spinel

In addition to the ruby, two other major gems, spinel and pyrope, owe their richer reds to chromium. In neither of these are the fine 'hair lines' in the red or in the blue regions of the spectrum so clearly or so consistently seen as they are in ruby. None the less the spectroscope can usually identify each with certainty.

It is well known that the finer reds and pinks of spinel under ultra-violet light or between suitable 'crossed filters' may show a fluorescent glow which can be as striking and as vivid as that displayed by ruby. But it should also be borne in mind that not all red spinels fluoresce in this way. To the unaided eye those that do, do so indistinguishably from ruby, as nearly the same wavelengths are involved in each case; but the spectroscope reveals a clear-cut difference in the patterns formed by the sharp emission lines which can usually be recognized at a glance without recourse to wavelength measurement of any kind. In spinel, unlike ruby, there are no strong absorption lines in the blue to act as a further aid to identification, so that an ability to recognize the characteristic grouping of red emission lines has an added importance.

In Becquerel's famous paper[1] of 1859 – a paper amazing in its originality and accuracy considering that spectroscopy was in its infancy at that time – the emission spectrum of red spinel under the influence of sunlight, as seen through Becquerel's phosphoroscope, had already been described and delineated, the positions of the lines being very accurately drawn in relation to lines in the Fraunhofer spectrum. Since then the spinel spectrum seems to have received little attention: Wherry[2] mentions only the broad central absorption, giving it an incorrect centre at 5600A (560nm) and attributing it wrongly to manganese. Moir[3] looked in vain for 'hair lines' in red spinel after observing them in ruby and in emerald. Even Deutschbein[4], who made detailed measurements of the fluorescent spectrum of spinel, could not find any corresponding absorption lines.

The emission lines of spinel provide an excellent exercise in spectroscopy, but observations are made far simpler if the light has first been filtered through a flask of copper sulphate solution, or similar blue filter, in order to see the bright lines standing out clearly against a dark background. In ruby over 95 per cent of the fluorescence is concentrated in the brilliant doublet at 692.8/694.2nm, with subsidiary lines playing a hardly noticeable part, in spinel a whole group of lines like a set of organ-pipes can be seen, diminishing in strength from the centre outwards.

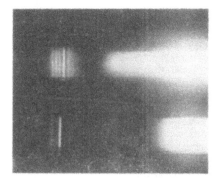 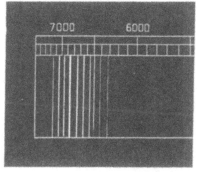

Figs 12.1 and 12.2 Typical 'organ pipe' fluorescent lines in RED SPINEL illustrated by a photograph in 12.1 and a diagram by T.H. Smith in 12.2. The lower image in 12.1 is a correctly aligned ruby spectrum for comparison

If the light is intense and the focus and slit-width of the spectroscope properly adjusted, a determined and critical eye should be able to perceive no less than ten narrow lines in ruby-red spinel, even through a small hand spectroscope. The strongest of them is at 686nm and this is clearly separated from the next most prominent line at 675nm by a dark gap. On the long-wave side there are three others in diminishing strength, while the short-wave side has a diminishing series of five lines beyond the 675nm already mentioned.

Fig. 12.1 is a photograph of this effect taken by us through a copper sulphate filter, showing the organ pipe effect in fluorescent spinel compared with the single bright line seen in a ruby. Fig. 12.2 is a scraper-board diagram of the spinel emission spectrum made by T.H. Smith for the original series of papers. The lines them-selves are about 2nm in width and most centres are separated by about 10nm. The actual wavelengths are difficult to measure despite their apparent sharpness.

The following values, the mean of many observations made with our full range of instruments, are accurate to within a nanometer. That for the 686nm line was checked against the barium emission line at 686.6nm using the table spectrometer.

Becquerel's drawing, it may be said, shows this line to be almost coincident with Fraunhofer's B line, usually given as 687nm. This also was checked by the authors when a glimpse of sunshine in overcast London gave them a chance to do so.

Emission spectrum of red spinel

716nm   w.
705nm   m.w.
697nm   m.
686nm   S.
675nm   m.s.
665nm   m.
656nm   m.w.
650nm   w.
642nm   v.w.
632nm   v.w.

656nm is noticeably narrower and sharper than the others – possibly because it is indeed a single line whereas the others, according to Deutschbein (who used a spectrograph of consider-able resolving power), are really close-spaced triplets, the compo-nents of which merge to look like single lines. The description of this fluorescence has been given in some detail and the reader urged to study it closely, but for identification purposes it is suffi-cient for the practised observer to have a glimpse of the two most prominent lines with the dark gap between them and a hint of others to either side to know with certainty that he is dealing with natural ruby-red spinel.

We have found this test on the luminescence spectrum of such spinels a most valuable check in dealing with quite tiny stones picked out as suspect under the microscope from parcels of cali-bre rubies. A refractometer test on such sizes would be difficult and tedious, and hardly as immediately decisive.

As with ruby, spinel is most brightly fluorescent when the

chromium activator is present in quite small amounts. Effects seen in deeper red stones are not nearly so striking. A group of red-pink octahedra from Burma glowed much more brightly between crossed filters of copper sulphate and spectrum red than another group of crystals with a deeper red colour, which indeed showed distinct absorption lines in their transmission spectra, an effect unobtainable in the paler stones.

Absorption lines

Although the absorption lines tally almost exactly with the red emission lines, as they do in ruby, the main absorption doublet does not actually coincide with either of the strongest emission lines. Our measurements gave 685.5 and 684nm for the doublet, which is far stronger than any others of the extremely feeble absorption lines at 675, 665, 656, 650, 642 and 632nm which correspond exactly with the emission lines already listed. In a number of chrome-rich (darker) stones weak lines have been seen in the blue at 465 and 455nm, and in one stone only, a line at 421nm. Vague traces of other lines have been seen in the blue but were not measurable (Fig. 12.3).

As with all red stones, there is a broad absorption in the green which in ruby-red spinel can be of great intensity. The position of this band is important in distinguishing spinel from pyrope garnet, which will be discussed in the next chapter.

Synthetic red spinel

The emission spectrum of red spinel has been found to show interesting variations from the normal 'organ-pipe' appearance, particularly in the synthetic versions of the stone. These differences were noticed and reported independently by both Gübelin[6] and Anderson.[7] It is well known that to grow large boules of spinel successfully it is necessary that they have a considerable excess of alumina; 1:3.5 for the MgO: Al_2O_3 ratio being commonly employed, the excess alumina being present, not as 'alpha' alumina, i.e. corundum, but in the unstable 'gamma' form which is cubic and isomorphous with spinel. Chromium in this gamma-alumina lattice produces a green colour and not red. However small red boules with the normal ratio of Mg and Al found in the natural stone can be grown, but with some difficulty. But these are unlikely to be popular since they offer no advantage over the much cheaper synthetic ruby, which they resemble closely, even to

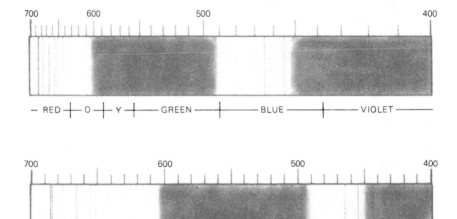

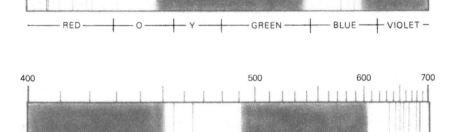

Fig. 12.3 RED SPINEL. The absorption lines of this stone are often faint and less diagnostic than the multiple emission lines of the finer reds. But the position of the broad absorption close to the red is useful to distinguish spinel from pyrope

showing bubbles and curved lines of growth. Samples of such Verneuil red spinels show a single line fluorescent spectrum which at first sight has much more in common with the ruby spectrum than with the 'organ-pipe' lines of natural red spinel. But the variation from natural spinel is more apparent than real and the effect is due largely to a strengthening of the 686nm line at the expense of others of the complex. In recent years other synthetic spinels have been made by crystallizing from a melt, one of the first of which was described in detail by R. Crowningshield and R.J. Holmes, also noting the ruby-like nature of the spectrum. These

90

nicely crystallized octahedral crystals have higher density and refractive index than normal and the latter may show variations within the one crystal.

Unfortunately it is not only synthetic red spinels which show emission lines having a ruby-like appearance: several natural reds have been seen in which the 686nm line is sufficiently dominant to make an error of judgement possible. However nothing resembling the three strong and characteristically grouped blue absorption lines of ruby are to be seen in any red spinel, so that their absence should serve to prevent error in the very few cases where the spinel emission spectrum is non-typical.

One other anomaly of synthetic spinel should be mentioned, if only because it was first noted by the late Robert Webster more than fifty years ago. He found that certain deep cobalt blue synthetic spinels fluoresced red and gave an emission spectrum remarkably similar to that of fine ruby. The massive block of absorbed wavelengths due to the cobalt was sufficiently like the central broad absorption of ruby to further the similarity. Apart from the totally different colours of the stones, careful measurement showed that the bright line was in fact two or three nanometers below the wavelengths of the ruby line, while the block absorption usually showed some indication of its composite structure, which could never happen in ruby.

13   Absorption Spectra of Pyrope and Topaz

When we speak of the absorption spectrum of pyrope we are really referring only to the chrome-rich forms of this garnet, which are the most used in jewellery. 'Pure' pyrope is almost unknown, but would be colourless since magnesium and aluminium cannot of themselves give rise to colour. All red pyropes so far known have contained substantial amounts of the almandine molecule, and this in itself would bring with it a red colour, due to the iron component, and the typical absorption spectrum of almandine with its three main bands at 575, 527 and 505nm and subsidiary bands, which will be described later.

We do in fact occasionally meet with red garnets which, from their low refractive index and density, must be classed as pyropes, yet which show a rather weak almandine spectrum and nothing more. One exceptional specimen of this kind, weighing 9.68 carats, was found to have the unusually low refractive index of 1.733 and density 3.670.

But in the rich red stones from Kimberley, Arizona, and Bohemia there is enough chromic oxide present to make a profound difference to the appearance of the spectrum, though the almandine bands are not absent but merely masked to a great extent by the overriding absorption effects due to chromium.

The absorption lines and bands to be seen in such chrome-rich pyropes are soon told. The usual narrow chromium doublet in the deep red is at 687 and 685nm, and is seldom strong. Weaker lines may be discerned at 671 and 650nm in some stones. There is a broad absorption nearly 100nm in width which is centred at about 570nm, and this covers two of the prominent almandine bands. The remaining strong almandine band at 505nm can, however, almost always be seen. There is general absorption of the violet beyond about 440nm (See Figure 13.1)

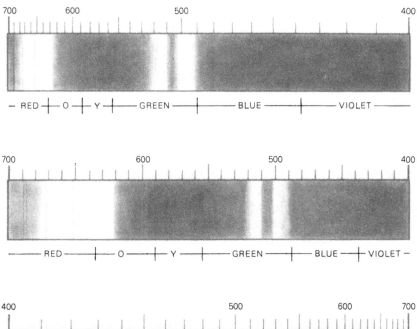

Fig 13.1 PYROPE GARNET. Again a faint pattern of chromium lines with a good broad absorption of the orange, yellow and green, further from the red than in spinel. The 505nm band of almandine is usually visible

Distinction from spinel

The general similarity in appearance between the spectra of pyrope and of red spinel was mentioned in the last chapter.

1. Pyrope always contains considerable iron, and this prohibits fluorescence. The luminescent spectrum of fine red spinel, on the other hand, is its most characteristic and constant feature.
2. While the narrow chromium lines in the red are difficult to observe in both minerals and are in any case too similar to act as useful identification, the broad absorption in the green is

93

always present but in different positions in the two cases. In pyrope it is centred near 570nm and in red spinel near 540nm.

3. Although the 576nm and 527nm almandine bands cannot be seen in a chrome-rich pyrope owing to the masking effect of the broad chromium absorption band, the 505nm almandine band can generally be plainly detected, and its wavelength is easily estimated without measurement as it is situated at the junction between green and blue in the spectrum.

Pink topaz

The sherry-coloured crystals of topaz from Ouro Preto in the south of Minas Gerais, Brazil, contain a trace of chromium. When these crystals are suitably heat-treated, the chromium takes its place in the crystal lattice, and a pink colour results. This gives rise to a chromium absorption and fluorescence spectrum, albeit an extremely feeble one. The main doublet is situated at about 682nm, and can be seen either as a faint absorption line or, rather more readily, as a fluorescence line by scattered light – more especially if this is passed through a suitable blue filter. This spectrum of topaz has very small diagnostic value, but it is rather important to know that a fluorescent line *can* be detected in pink topaz under favourable circumstances, since a beginner might possibly think that the stone he was testing must be pink sapphire when he sees this line. The fluorescent doublet in pink corundum is incomparably more intense, and usually the three ruby absorption lines in the blue will also be visible. But if there is any doubt in the mind of the observer a refractive index or other check test is advisable.

Some practical guides to wavelength

In advocating the use of the spectroscope as a means of gem identification we have always maintained that the work can be done with the simplest form of spectroscope and without wavelength measurements. There is nothing to be said *against* such measurements, of course; but they do entail more expensive apparatus and a greater expenditure of time. Also, unless the wavelength scales are accurate, they may tend to mislead the observer who depends too much on them.

There are a few quite simple wavelength guides, however, which may on occasion be useful (apart from the spectrum colours themselves, which should enable one to make an estimate within ten or twenty nanometers) and it may be appropriate to mention them here.

1. Where sunlight is available, this can be used both as the contin-uous source and also as a wavelength guide, since the main Fraunhofer lines are fairly easily recognizable, and their wave-lengths obtainable from the literature. The B line at 687nm can just be discerned in the deep red; the C line in the orange-red at 653.5nm, the famous D lines at a mean wavelength of 589.3, E at 527 in the green, and F at the beginning of the blue at 486.1nm will be found the most useful. To the blue side of E is a group of magnesium lines designated as 'b' which are also rather prominent; 518nm can be taken as the mean wavelength for this group.
2. The emission lines of sodium or of mercury vapour are often available to the gemmologist, and it is easy to arrange matters so that these bright lines can be seen superimposed on the light from the continuous source used for seeing absorption bands. The mean wavelength of the sodium doublet is of course 589.3nm while the mercury lines, which can be seen in 'daylight' fluorescent tubes are most conveniently spaced in the yellow at 578, the green at 546 and blue at 438nm. (These wavelengths are taken from modern tubes and differ from those quoted by Anderson, so it must be assumed that the contained gases are varied from time to time, and it may be necessary to adjust one's assessment of approximate wave-lengths by reference to the colours of the bright lines.)
3. Crookes glass, used for anti-glare spectacles, and for furnace workers' goggles to absorb the incandescent yellow light of sodium emission, contains didymium (neodymium) and has two main blocks of absorption lines, the strongest of which is at 584nm. This can be used to determine whether a broad absorp-tion band is centred at 570nm (pyrope) or at 540nm (spinel). If such spectacles are still made today, students wearing them may be puzzled by a persistent absorption of the yellow seen in every stone tested by spectroscope. The explanation for this is obvious, and it is only necessary to take the faintly tinted glasses off to get a normal spectrum.

Another useful group of absorption bands for comparison is provided by that cheap and readily obtainable chemical, potas-sium permanganate. A few small crystals of this in a small glass dish of water will make a pale purple solution giving five bands

centred at 571, 546, 524, 504 and 486nm, and if the red spinel or
pyrope is dropped into this, its absorption will be seen against the
permanganate spectrum, again enabling the approximate wavelength of the tell-tale broad absorption to be determined. Figs.
13.2 and 3 illustrate these two aids.

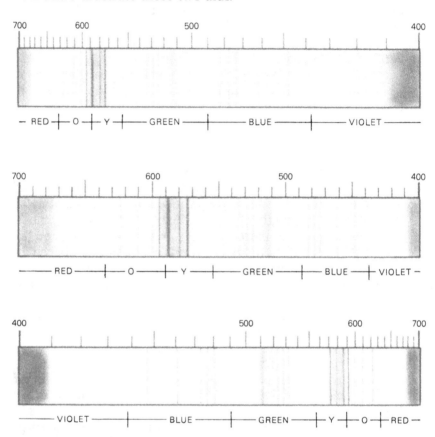

Fig. 13.2 DIDYMIUM GLASS. A useful spectrum from which to assess
approximate wavelengths of other spectra

A further aid might be found in the spectrum of the Chelsea
emerald filter, since this passes only two narrow bands of light,
one in the deep red, from 690 to 715nm and the other, the brighter
one, since the eye is most sensitive to green, from 560 to 580nm. If
this filter is passed momentarily in front of the eye while looking
at a spectrum from a stone the two bands can be related to that
spectrum and may help determine approximate wavelengths.

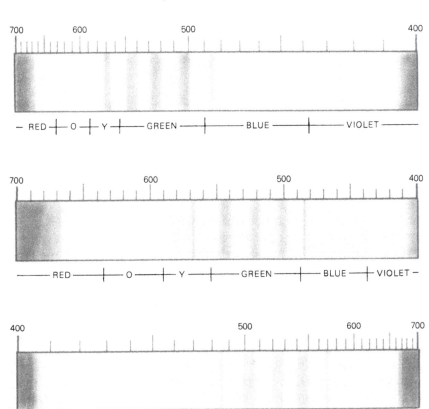

Fig. 13.3 POTASSIUM PERMANGANATE SOL. Another very useful comparison spectrum for assessing wavelengths of other spectra. Only a weak solution is necessary

14   Absorption Spectrum of Emerald

In the range of green stones which owe their colour to chromium, emerald plays the same predominant role as ruby in the red series, both in the magnificence of its colour and in the sharpness and invariability of its absorption spectrum. As has already been suggested, the two factors are not unconnected. A mineral having a clear-cut absorption spectrum in which bands have well-defined limits and colours transmitted are not obscured is far more likely to display a brilliant, clear residual colour than one in which there is an all-over slight absorption with no definite bands.

Despite the change-over from red to green, anyone familiar with the ruby spectrum can recognize the general similarity of this to the absorption spectrum of emerald, and realize that the same colouring ion is responsible for both. The strong, narrow doublet in the deep red, followed by less prominent narrow bands on the orange side, a general absorption region in the yellow-green, narrow lines in the blue, and general absorption of the violet – all these, in general terms, are common to both.

The main reason for the striking difference in colour is the great divergence between the position, width and strength of the central absorption region. In ruby, as we have seen, this spreads over the whole of the yellow and green (roughly 600 to 500 nanometers). In emerald it extends only from about 625 to 580nm, and is far less intense. Thus, though the deep red is almost as freely transmitted as in ruby, the whole of the green is also a transparent region, while in ruby this is obscured.

We shall see later how alexandrite – another of the chromium clan – has its central absorption band in an intermediate position, thus accounting for its well-known apparent change in colour when viewed in daylight or in more yellowish artificial light.

Sir Lawrence Bragg suggested that in ruby and spinel the

chromium ion is held tightly and chokingly, giving 'red-in-the-face' colours, whereas in emerald there is more space in which the ion can be accommodated and a green colour results. In the case of alexandrite there is a balance between the tight squeeze of a red stone and the easier situation in a green one and the colour changes with the nature of the light source. The reasoning here was not scientific, but served to impress the facts on the mind of the student. Sir Lawrence was a brilliant lecturer, able to adapt to each type of audience.

J. Moir[1] was the first to observe and record the 'hair lines' in the red part of the emerald spectrum. It has not received nearly so much attention as the spectrum of ruby, nor has it been previously described with any completeness. The present authors were apparently the first to notice and to measure the narrow bands in the blue region which are described here. Deutschbein,[2] in the great paper already cited, gives some accurate measurements for the red luminescence and for the red absorption lines. Our own measurements for the doublet and the main blue line were checked by using a diffraction grating on a table spectrometer and comparing with suitable emission lines of strontium (679.1), manganese (478.3), and zinc (481nm).

Consider the emerald spectrum in greater detail. Even more than in ruby, there is a marked difference between the spectrum of the ordinary and extraordinary rays – enough difference, in fact, to be worth noting even by those whose interest in absorption spectra is purely practical.

In the spectrum of the ordinary ray only two narrow lines can be seen in the red, and these are of almost equal strength. One consists of the main doublet, which is consistently present in all chromium spectra, the components in this case having wavelengths 683 and 680nm, and the other is a clear-cut line at 637nm. There is a rather weak and diffuse central absorption band from about 625 to 580nm, as already mentioned, and a narrow line in the blue at 477.5nm, which is clearly seen only in chrome-rich specimens under good lighting conditions. A weaker line at 472.5nm has also been observed in a few specimens. From about 460nm onwards the usual chromium absorption of the violet commences.

In the 'extraordinary' spectrum the doublet is rather stronger, particularly the 683 component. In place of the 637 line, which is missing, two rather diffuse lines at 662 and 646nm are noticeable features. These are bordered on the shortwave side by narrow regions of extreme transparency, giving a curious and characteristic appearance. The broad central absorption band is so faint as to be hardly noticeable, but its shadow extends further towards the red than it does in the ordinary ray, and there are no lines in the blue.

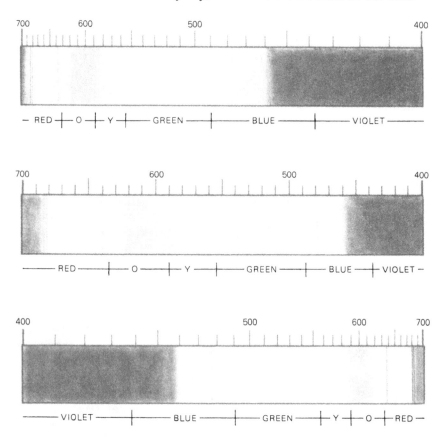

Figs 14.1 EMERALD. This is the absorption pattern of the ordinary ray

It must be realized that, unless light due to the ordinary ray is cut off by interposing a polarizer between the stone and the spectroscope, the 'pure' extraordinary spectrum can never be observed, as the ordinary ray must always represent at least 50 per cent of the light reaching the slit. On the other hand if, as often happens, an emerald is cut with its table at right angles to the optic axis the spectrum of light passing through this facet will be virtually 100 per cent due to the ordinary ray. Thus, when an observer sees the 'ordinary' spectrum only, with its two narrow lines of equal strength in the red, he knows that the stone has been cut in this favoured orientation and the blue-green appearance of the stone as viewed from the front will bear this out. On the refractometer, a stone cut in this way will show full double refraction however the stone is turned on the instrument.

If the emerald, table facet up, displays a more yellowish or jade-like green colour, it is a sign that it has been cut in such a fashion that the extraordinary ray is playing a notable part in the light emerging from the stone in this direction. To some eyes this is more pleasing than the deeper tint of the ordinary ray. The 637 line will then be less intense compared with the doublet, and the two diffuse bands between these will be prominent.

The absorption spectrum of the ordinary and extraordinary rays are shown in Figs. 14.1 and 14.2

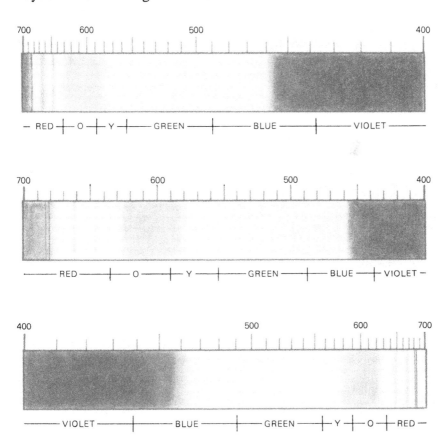

Fig. 14.2 EMERALD. The absorption pattern of the extraordinary ray. Polaroid is needed to separate these two spectra and without it emeralds may give a combination of the two in both natural and synthetic stones

The strength of all these lines and bands will naturally vary with the quantity of chromic oxide present. Synthetic emeralds are particularly chrome-rich, and thus show a very strong spectrum. Indeed, when the first German 'Igmeralds' appeared before the war the authors noticed two 'extra' bands belonging to the extraordinary ray. These were in the orange at 606 and 594nm, and we at first hoped that they were diagnostic for the synthetic stone.[3] But natural Colombian emeralds of exceptionally deep hue may sometimes show the bands, while many synthetics do not, so their presence cannot be regarded as diagnostic of real or of synthetic emerald.

The absorption spectrum of emerald is sufficiently characteristic to make it of great practical importance in recognizing this variety of beryl. There are many other shades of green beryl, but emerald is the only variety to show chromium lines. Thus, however pale the tint, where the chromium lines are seen in a beryl the stone is entitled to the name of emerald.

The same useful criterion can be made to apply in the case of alexandrite. Fluorescence can be made to form another test, which will be mentioned later.

Where emeralds are mounted in closed settings and backed with green-coloured foil to enhance the apparent depth of colour of the stone, as was often the case in antique jewellery, the strength of the absorption lines will give a useful clue to the true depth of colour of the stone itself, for colour applied or reflected from outside the stone will have no effect whatever on the intensity of the absorption.

Though several other green stones owe their colour in part at least to chromium and show a chromium spectrum, either the appearance of the spectrum or of the stones themselves is so different from that of emerald that there should be no confusion to an experienced observer. Green fluorspar and certain heat-treated indicolite tourmalines are the natural stones which most closely resemble transparent emerald, while quartz or synthetic spinel 'soudé' emeralds, or green pastes are the most common imitations, and none of these show a chromium spectrum except a few glasses, and in these the bands are so woolly that even a tyro could hardly make a mistake.

Jadeite sometimes has a fine emerald-green colour, and this is due to chromium and gives rise to an absorption spectrum not very unlike emerald, so that to separate fine translucent jadeite from translucent forms of emerald (which admittedly are seldom seen in jewellery) does need a certain degree of care if the spectroscope alone is used as arbiter. Apart from wavelength differences, the jadeite spectrum is not nearly so well defined as that of emerald, and a polaroid or nicol will not make that striking difference to the spectrum when turned before the spectroscope slit as we have described above in emerald. Further, jadeite has a powerful and distinctive band of its own in the violet at 437nm, which can usually be seen and is completely diagnostic. This leaves aside all those differences in texture, lustre, refractive index, specific gravity, etc., which can be called to one's aid if need be, let alone a simple test with a Chelsea colour filter.

Fluorescence and the Chelsea filter

Mention of the Chelsea filter reminds one that people often have trouble with this little aid to colour analysis because they do not understand its working or its limitations. Essentially it is a very efficient dichromatic filter – that is, it transmits two narrow regions of the spectrum only: one in the deep red and the other in the yellow-green. Through such a filter materials may appear red, yellow-green, or a rather brownish indeterminate shade compounded of the two. It so happens that emerald, though a green stone, transmits fairly freely in the deep red, and absorbs a good deal in the yellow-green. For this reason it tends to show red under the Chelsea filter, while green glass, most soudés, green tourmaline, and green jade, absorb the deep red and appear green through the filter. Unfortunately, emerald-green fluorspar, demantoid garnet, green zircons, and certain exceptional soudés, appear reddish under the filter, while genuine emeralds from the Transvaal and from the Jaipur district in India show little or no red effect. Moreover, synthetic emerald can appear even more red through the filter than the natural stone.

Observations of the fluorescence of emeralds from different localities, using the crossed filter method,[4] convinced us that the red appearance of emerald through the Chelsea filter was largely due to the degree of fluorescence emitted by the stone when placed in a strong white light. Synthetic emerald which shows so striking a fluorescence between crossed filters, also shows an intense red glow through the emerald filter, and in natural emeralds a diminishing fluorescence is paralleled by a lessening of the red seen through the filter.

However, a close study of the spectrum of light reflected from a fluorescent emerald shows that the effect seen through the Chelsea filter is enhanced by the transmission of further wavelengths in the deep red which correspond with the red transmission region covered by the filter. But in all emeralds the transmission of green wavelengths in ordinary light far exceeds that of the red wavelengths, so that the gem looks green. It is only when the narrowly selective wavelengths seen through the Chelsea filter are examined that we see the red residual effect.

South African and Indian emeralds which are sometimes said in textbooks to be green through the filter, are in fact better described as ash-grey with scarcely a hint of either pink or of green, and are also non-fluorescent because they contain some iron, the great killer of luminescent effects. Any 'emerald' that looks really green through the Chelsea filter is almost certainly not an emerald. Used with discretion the filter can be of great service to the jeweller, but it can be understood from the foregoing remarks how difficult it is to cover all contingencies in a brief set of directions issued with the filter unless the user can have some glimmering of the reason for the different behaviour of the gems which he wishes to differentiate. There are further complications, too, when it comes to blue stones; but this is not the place to discuss the filter except in its relations to the spectral peculiarities of emerald.

Summary

Emeralds, from whatever locality, show a very distinctive spectrum, typical, in its general character, of spectra due to chromic-oxide to which the magnificent colour of the gemstone is undoubtedly due. The main doublet is at 683 and 680nm, and can be resolved into two lines by even a small diffraction grating spectroscope. Two more diffuse bands on the orange side of the doublet are at 662 and 646nm. These have a peculiar appearance owing to their being flanked by regions of great transparency on their short-wave side. There is a narrow clear-cut band at 637nm, which, in the 'ordinary' spectrum, is as strong as the doublet, and is the only other red line besides the doublet which is seen when looking down the optic axis of the stone. There is a not very noticeable general absorption of the yellow-green, and a narrow band of fair strength at 477.5nm belonging to the ordinary ray. In deep coloured stones, including synthetics, bands at 606 and 594nm may appear in the spectrum of the extraordinary ray. A feeble line at 472.5nm has sometimes been observed in the ordinary ray spectrum. Apart from synthetic emerald, which can be distinguished by a variety of other methods, green jadeite is the only stone which a careless worker might confuse with emerald on the basis of its absorption spectrum.

15   Absorption Spectrum of Alexandrite

Since the daylight colour of alexandrite is usually some shade of green, we can properly include it in the series of green gemstones owing their colour to chromium: but the balance of absorption in its spectrum is so precarious that in incandescent light the green gives place to a reddish hue, and this 'colour-change' effect is further complicated by the strong pleochroism exhibited by this strange gemstone. In a good specimen of alexandrite there are in fact three distinct colours, which can only be separated by means of a dichroscope, polaroid, or nicol prism, but which cause considerable variations in the appearance of the stone when viewed from different directions. From the same piece of rough, indeed, two stones could be cut which would seem to have quite different colours.

The ray corresponding to the highest refractive index (gamma) in alexandrite is green, while that corresponding to the lowest index (alpha) is red or purple. The intermediate ray, corresponding to the 'beta' refractive index, is orange in tint. So far as the absorption spectrum is concerned, we have only to concern ourselves with the green and purple rays, since the absorption effects in the orange ray are feeble, and contribute nothing to the absorption bands seen in random orientations of the stone. In both the green and the purple rays the absorption spectrum is quite obviously that of a chromium-coloured mineral. Narrow, well-defined lines in the red, a broad absorption region in the middle of the spectrum, a narrow line or lines in the blue, and complete general absorption of the violet – all these well-known features are present. In details, however, the spectra of the two rays are sufficiently different to merit a separate description. Figs. 15.1 and 15.2 illustrate these.

In the green ray the narrow doublet at 680.5 and 678.5nm is very prominent, 680.5 being the stronger of the two. Weaker lines may be seen at 665, 655, 649, and 645nm. The broad central absorption is from about 640 to 555nm and seems to have two deeper concentrations within it, though these are too ill-defined to measure. There is complete absorption of the blue and violet from about 470nm (Fig. 15.2).

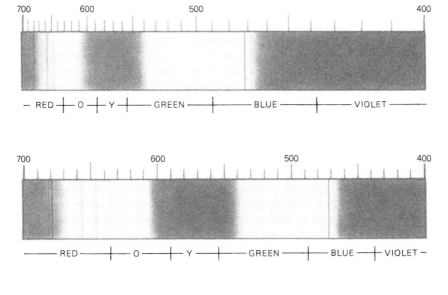

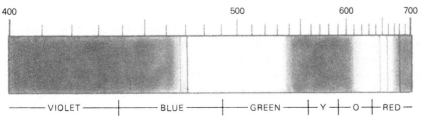

Fig. 15.1 ALEXANDRITE. In the red ray the shift is towards the blue, covering the yellow-green and allowing much more red through, so the colour tends to red

In the red or purple ray the doublet is weaker and 678.5nm is now the stronger of the two lines. Only two other lines are seen in the red; 655 and 645nm. The broad absorption has shifted well away from the red, at 605 to 540nm, accounting for the change of colour, and there is a clear-cut line in the blue at 472, sometimes with a weaker one at 468nm. The absorption of the blue has receded a little to 460nm (Fig. 15.1).

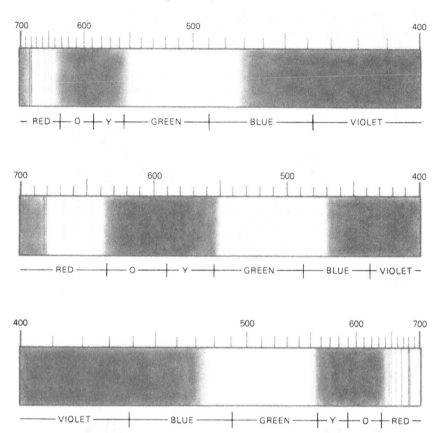

Fig. 15.2 ALEXANDRITE. In the gamma (green) ray the broad absorption is nearer the red, leaving the green transmitted, so the stone looks green

The above measurements were carried out on specimens of Siberian alexandrite, in which there is more chromium than in the Sri Lankan stones, which occur in larger sizes and are less flawed. The narrow absorption lines will be seen in the same positions and relative strengths in stones from either locality, but the general absorption regions are usually less intense and less extensive in the alexandrites from Sri Lanka, giving the stones a tendency to brownish-green daylight colour with a similar modification in incandescent light.

Unless a nicol or polaroid is used, the absorption spectrum seen in normal practice will of course be a 'mixed' one, and will vary slightly if the stone is viewed in different directions. It will always be noticeably a chromium spectrum, however, and there is no stone in the least resembling alexandrite which has at all a similar spectrum. If one is working by *artificial light only* it is possible to mistake a fine Siberian alexandrite for a Thai ruby on casual inspection, and with 'Thai ruby' in one's mind a quick glance at the obviously chromium-type spectrum might confirm, or seem to confirm, that error. This may sound far-fetched, but an error of this kind was actually made in the experience of the authors, and was only discovered when the stone was seen to be green in daylight. Although it may be acceded that the spectrum of alexandrite and of Thai ruby are rather similar by direct transmitted light, when the ruby doublet may show as an absorption line and not as a bright fluorescent line, inspection of the blue region should remove any doubt. The three strong lines in the blue are always prominent in ruby, and their 'pattern' of two close together and one well separated is also distinctive. In alexandrite it is difficult enough to see even *one* line in the blue, so that mere observation in this region is quite sufficient to differentiate it from ruby.

Fluorescence in alexandrite

Nothing has yet been said about the 'reversibility' of the doublet and other lines in the red of alexandrite, corresponding to the effects seen in ruby and to some extent in red spinel and in emerald. In alexandrite a similar effect, due to fluorescence *can* be seen, but it is hardly likely to be noticed under ordinary conditions, using transmitted light. Under 'crossed filters' all alexandrites can be seen to glow distinctly red, and 'copper sulphate light' transmitted through the stone and examined with a spectroscope will show the doublet as a fluorescence line, with a rather vague fluorescence patch to the orange side of it. Rather curiously, it is in Sri Lankan alexandrites, containing relatively little chromium, that the red fluorescence is most strongly seen, and it forms a most sensitive test for the presence of chromium in chrysoberyl, without which it cannot properly be described as 'alexandrite'.

Deutschbein[1] gives a detailed description of the emission and absorption spectrum of alexandrite, including several more lines that we have been able to observe; but he takes no account of the pronounced variations due to pleochroism. There seem to be few other references in the literature to the alexandrite spectrum: Wherry[2] includes it in his comprehensive paper, but he gives a very incomplete description and makes the mistake of ascribing the bands to vanadium.

Summary

Alexandrite owes its well-known 'colour-change' and striking pleochroism to the presence of chromium. Any chrysoberyl, therefore, which shows no trace of chromium absorption lines nor a red fluorescence between suitable crossed filters cannot be classed as an alexandrite.

The spectrum is unmistakably a chromium one, including a strong doublet in the deep red and several weaker lines on the orange side of this, as well as general absorption in the yellow region and in the violet. A single narrow line can usually be detected in the blue, just before the general absorption sets in. The lines in the red are unusually sharp and narrow: those in the orange-red show none of the peculiar diffuse effect seen in emerald with its transparency patches alongside, while the difficulty with which the doublet can be seen as an emission line distinguishes it from ruby, in addition to the lack of the three very characteristic ruby lines in the blue. Indeed, to the practised eye, the spectrum of alexandrite is completely diagnostic, especially when the general appearance of the stone is taken into account.

When observing the spectrum of light which has been transmitted in different directions in the stone, variations can be noticed in the relative strength of the narrow lines and in the position of the central absorption region. By using a polaroid disc, the spectrum due to the green ray and to the red ray can be studied separately if so desired.

16 Absorption Spectrum of Jadeite

Many green stones in addition to emerald and alexandrite show chromium lines in their absorption spectra, but only two, green jadeite and the green garnet, demantoid, are of commercial importance in jewellery. These two gems are alike in having more powerful and more diagnostic absorption bands in the violet region which are due not to chromium but to iron. Since the fine colour of each is so largely due to chromium their spectra will be described with others of the chromium group, back-references being made when the time comes to discuss iron spectra.

At its best, jadeite from its only important source in Upper Burma has a colour as fine as that of emerald – and indeed so closely resembling it that only its lack of transparency and its texture prevent confusion between the two. Fine pieces of jadeite exhibit a considerable degree of translucency, and quite enough light can be transmitted from a 250- or 500-watt source to enable the typical chromium absorption bands to be seen with a prism spectroscope. Were it possible to view the spectrum from a single large crystal of the jadeite mineral there is little doubt that a series of clear-cut lines would be seen very similar to those in emerald, and showing directional differences due to dichroism. But we invariably encounter jadeite in a fibrous or granular form, so that the spectrum seen is an average one subscribed to by a large number of grains in different orientations. It is thus, as one might expect, a little blurred as compared with single-crystal chromium spectra, and remains uninfluenced by direction or by the interposition of a polarizing filter. The strongest band is, as usual, in the deep red, and is centred near 691nm. This is probably an unresolved doublet with 'limbs' at 694 and 689nm. This main band is accompanied by a rather weaker one at 655nm and a still weaker and vaguer band at 630nm. These bands have a similarity to certain of the emerald bands in having a transparency patch alongside them. They increase in strength with the chromium

111

content and depth of colour. When strongly represented there is fairly deep general absorption of the blue as well as of the violet region, and no further bands are visible. In the paler green specimens and in thin, very translucent pieces of the deep green types, further bands may be seen in the violet end of the spectrum. By far the most important of these, on account of its intensity and its almost universal presence in all specimens of jadeite, whatever their colour, is a rather narrow band in the violet at 437.5nm. This wavelength was checked pretty accurately by reference to the two barium emission lines at 440.2 and 435nm which could be seen to be almost equidistant above and below the absorption band in question. Accompanying the strong absorption band are similar but much weaker bands at 450 and 433nm, and more rarely a vague band near 495nm has been seen. The pale and very translucent 'water jade' shows the violet bands at their clearest, but the 437.5 band can be detected in almost any pale jadeite by reflected light, and it forms a very valuable means of distinguishing this attractive and prized gem material from the many natural minerals, not to mention imitations, with which it can be confused.

One gem mineral does indeed show almost identical absorption bands in the violet to those of jadeite, but as its appearance is entirely different this should not be at all dangerous. The stone referred to is spodumene; not in its well-known kunzite variety, which shows no bands, but the yellow and green types, which contain traces of iron to which there can be no doubt the bands are due. It will be remembered that spodumene is closely analogous to jadeite in composition, being a lithium aluminium silicate while jadeite is a sodium aluminium silicate, and both are monoclinic pyroxenes; it is thus not surprising that their absorption spectra are similar.

Fig.16.1 shows the typical absorption of green jadeite in which the chromium lines dominate, while the strong 437nm band is seen only dimly through the general absorption of the violet. Fig. 16.2 gives the absorption of a light-coloured translucent jadeite of the type known as 'water jade', where no chromium lines are present, but the 437nm and its accompanying weaker bands are clearly seen.

'Yunnan Jade'

A coarse and rather ugly green type of jadeite is occasionally met with in which the chromium content is so great that the material is almost opaque. When light is passed through a thin piece of this, the spectroscope reveals virtually only one band of light – a patch

112

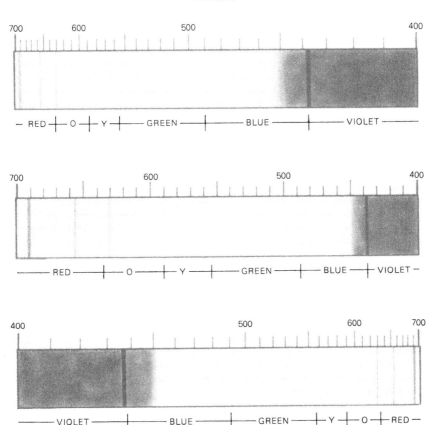

Fig. 16.1 GREEN JADEITE. Deep green specimens will show chromium lines in the red, but the most persistent band is at 437.5nm and is due to iron

of green, with its centre near 535nm – all other colours having been absorbed. With a very powerful beam of light and a very thin piece of this jade, two other 'windows' can be dimly discerned – one centred in the deep red near 695nm and the other in the orange near 620nm. The spectrum of this peculiar and unattractive form of jadeite is so apparently dissimilar to that of normal jade that we were at first reluctant to accept it as the same mineral, although the density and refractive index were correct enough for jadeite. An X-ray powder photograph proved beyond doubt, however, that the mineral was indeed jadeite. Some knowledge-able lovers of jade class this as 'Yunnan jade'; but since, according to Mr Howard Hansford, who made a special study of jade and its

origins, jadeite from the Tawmaw district of Upper Burma originally entered China in the eighteenth century through the
province of Yunnan, the name was probably attached to it for this
reason and not because it was mined in this part of China. It is
quite possible that its strange-looking spectrum is essentially the
same as that of normal jadeite, only very much exaggerated due to
the presence of an excess of chromium.

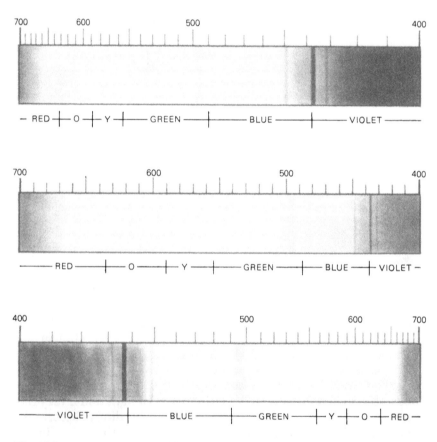

Fig. 16.2 'WATER JADEITE'. This transparent form of pale green jadeite may
show only the 437.5nm band and one or two fainter ones due to iron

Nephrite

That other jade mineral, nephrite, usually owes its green colour to
iron rather than to chromium; hence its sadder tints of spinach
green. Some of the finer green pieces, however, do contain traces

114

of chromium, with the strongest band in nearly the same position as in jadeite. But this is so vague and obscure that it can hardly be seen, and there are no bands in the violet remotely resembling the characteristic 437.5 band of jadeite.

Stained jadeite

In 1957 some fine green jadeites were found to have been stained, the original colour of the material having been white or grey. In addition to the fact that the dye-filled cracks could usually be detected by lens or microscope, these stones gave a pinkish red through the Chelsea emerald filter and showed a broad absorption band centred at 660nm and a weaker one at 600 (Fig. 16.3). Heavily stained jadeite cuts off completely at about 600nm and the jadeite band at 437nm may be masked by the absorption of the violet.

Apart from the dye-filled cracks, all these effects were also to be seen in hollow jadeite triplets with a dyestuff filling, which were marketed at about the same time, although their absorption bands were shifted about 20nm towards the blue. It should be mentioned that both the dyed stones and the dye-filled triplets are liable to fade if exposed to light too much. One triplet faded badly after twenty years, even though it was kept in the dark.

One instance is known of a thin hollow-backed cabochon of exceptionally fine dark green (Yunnan?) jadeite which showed an absorption band in almost the same place as these dyed stones. For a time this stone was confused with them, although it did not give the residual red or pink through the filter, and had no dye-filled cracks.

Some early stained or dyed mauve jadeites were found to have rather similar absorption bands. The G.I.A. have recently reported tests on some mauves which did not have this absorption. They did, however, show dye concentration in cracks.

Summary

The absorption spectrum of jadeite is sufficiently distinctive to enable the mineral to be identified in nearly all cases either by transmitted or reflected light. In view of its close resemblance to some varieties of nephrite and to massive grossular, chrysoprase, smithsonite, bowenite, and other jade-like minerals, the spectroscope test, which is often the only one which can be easily applied, is of first-class importance to the practising gemmologist.

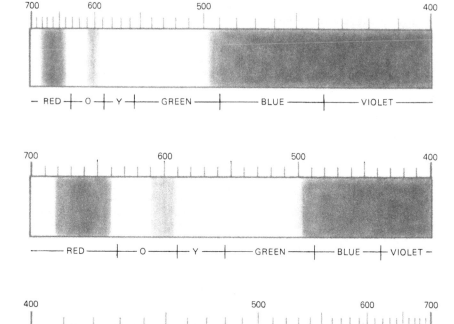

Fig. 16.3 STAINED GREEN JADEITE. This has a totally anomalous broad band centred at about 660nm and a fainter one at 600nm. These are due to a vegetable dye

In the case of medium to deep green jadeite the most clearly seen bands will be a typical chromium group of lines in the red. The strongest, near 691.5, is probably a doublet, in keeping with all other known chromium spectra; it is accompanied by weaker bands at 655 and 630nm. In specimens of pale green or of other pale colours a narrow band in the violet at 437.5nm is usually clearly visible, sometimes at great strength, often accompanied by weaker bands above and below this. No other jade-like mineral has a band at all resembling this.

The chromium spectrum of jadeite has a considerable resemblance to that of emerald, and since translucent emerald has a

116

similar appearance to jade on casual inspection a certain amount of caution is necessary. The emerald spectrum is more clear-cut; it also changes noticeably when a polaroid disc is rotated either below the spectroscope slit or above the eyepiece, while that of jadeite alters not at all under like conditions. Where the 437.5 band can be seen in the violet this forms conclusive proof that the stone is jadeite, since green or yellow spodumene is the only mineral with a similar spectrum. The 437.5 band is present in practically all colours of jadeite, though the general absorption in deeply coloured specimens may obscure it. It is helpful to concentrate the light through a blue filter, which effectively cuts out glare from the unwanted red or yellow parts of the spectrum, and the slit of the spectroscope can be widened considerably to admit more light, since nicety of detail is not necessary in such a case. Despite the presence of chromium, jadeite, unlike emerald, does not show red under the Chelsea filter nor between crossed filters. A jadeite showing pink or red under the filter is a stained stone.

17 Absorption Spectrum of Demantoid Garnet

Demantoid was the name coined by mineralogists for a green transparent form of andradite garnet which occurs in serpentine deposits in the Ural Mountains, and made its first appearance as a gemstone towards the end of last century. Amongst other even less fortunate names, such as 'Uralian emerald', with which it was dubbed by the trade, the name 'olivine' persisted, but with international rulings on names, and trades descriptions legislation in the UK, combined with the very considerable increase in the value of the fine green variety, it looks as if, in the 1980s, the name 'demantoid' was at last accepted. 'Olivine' was already attached by more than two centuries of use to quite another mineral, the magnesium-iron silicate known to jewellers as 'peridot'. The gemmologist, who realizes the importance of keeping the nomenclature of gems as clean and tidy as possible, will perform a useful service if, by force of example, he encourages the use of the name 'demantoid', which has quite a fine ring to it and can cause no confusion.

Andradite garnet is a silicate of calcium and iron, which can be represented by a formula of the usual garnet kind: $3CaO.Fe_2O_3.3SiO_2$. In nature it is often quite black, probably due to the intrusion of some ferrous iron in addition to the prevalent ferric form, since the presence of iron in two states of valency always tends to produce a dark colour. It is then known as 'melanite' or 'common garnet'. Of the transparent types, only the green demantoid has any place in jewellery. The name 'topazolite' has been suggested for a yellow form, but this is exceedingly rare and the name is best forgotten; should such a stone occur it can quite well be classed as a yellow andradite or yellow demantoid.

The andradite molecule can be replaced to some extent by those of other garnets, though it is usual for andradite to correspond much more closely to its theoretical formula than is the case with the other gem garnets. One may consider the chrome-green variety to contain a certain percentage of the chromium

118

garnet uvarovite ($3CaO.Cr_2O_3.3SiO_2$). Uvarovite itself, though rare, is occasionally found as masses of bright green, almost microscopic, dodecahedral crystals of a size that precludes cutting. However, lustrous crystals up to about 2 cm have been found at Outokumpo in Finland, but these are opaque and far too dark for gem use. A few may have been cut as collectors gems, but such sizes must have had greater value as crystal specimens.

Demantoid usually has a distinctive appearance on account of its bright green colour, its brilliant lustre, and its 'fire'. It is also typically seen in rather small sizes, so that any stone of over about five carats is much less likely to be a demantoid. However, one cannot of course always rely upon appearance only, and, since its refractive index is well above the range of the standard refractometer, it is fortunate that there are two easy tests by which its identity can be definitely proved. The first consists in an observation of its inclusions, which are more distinctive and more often visible than those in any other gemstone. Once the 'horsetail' bundles of fine asbestos fibres – usually radiating from a single point – are seen, there can be no confusion with any other green gem. The second conclusive test consists in its absorption spectrum.

Demantoid, as we have seen, contains ferric iron as an essential part of its composition, and to this must be ascribed the very intense absorption band at the beginning of the violet, with its centre at about 443nm, which is invariably present (Fig.17.1). In pale greenish-yellow specimens this band is sharply delineated and is not very broad. In the finer green specimens, in which the chromium absorption effects are superimposed on those due to iron, there is the usual strong general absorption of the violet, so that the 443nm band may be seen only as a sharp 'cut off' to the end of the spectrum. Widening the slit and using a blue filter may help to render the band visible in such cases. To compensate for this obscurity in the violet there are lines due to chromium to be seen in the red. Admittedly these are not so clearly visible as those in emerald or alexandrite, but on the other hand their unusual character and distribution helps to make them distinctive. There is a narrow band in the very deep red which we have not succeeded in measuring: the doublet (unresolved) is itself very deep into the red and is near 701nm. There is a weak but sharp line at 693nm, and two bands in the orange at 640 and 622nm, which are more easily seen. These two bands, unlike the usual chromium 'hair lines', are quite vague in outline and nearly 10nm broad. In addition to the powerful 443nm band, we have observed and measured weak bands at 485 and 464nm in the blue in certain fine specimens.

Demantoid is very transparent in the deep red, which accounts for its pinkish colour under the Chelsea filter, though it lacks the absorption of the yellow which would throw the emphasis more wholeheartedly on to the deep red component of the filter.

One may note that there is a certain parallel between the absorption spectra of jadeite and of demantoid in that the paler specimens show a strong and distinctive band in the violet due to iron, while in the finer green specimens this tends to be obscured by the presence of chromium which, in compensation, so to speak, causes the appearance of narrow lines in the red, which then aid in the identification of the mineral.

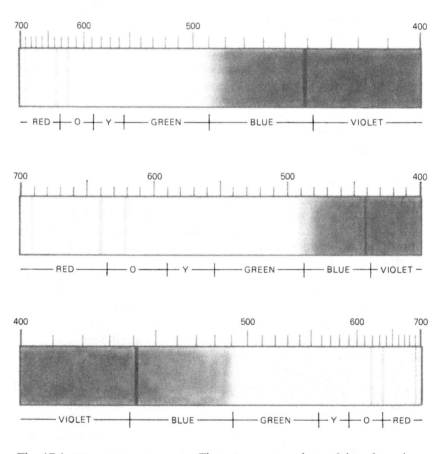

Fig. 17.1 DEMANTOID GARNET. Fine stones may have faint chromium lines in the red, but the intense absorption at 443nm is probably the best identifying band. It may be seen as a cut-off at the violet end of the spectrum

120

Gems similar to demantoid

Of the stones with which demantoid may be confused on grounds of appearance, only green zircon, sphene, and the very rare green diamond, give negative readings on the refractometer. Green zircon either gives a complete if rather blurred series of bands throughout the spectrum or, if of completely 'low' (metamict) type, will show a single vague band in the red and *no* absorption band in the violet. The strong double refraction (easily seen with a lens) and the marked dichroism of sphene will distinguish this remarkable gem without recourse to the spectroscope. As for green diamond: if it shows any spectrum at all it will be a narrow line at 504nm, and perhaps even fainter lines near this. Of other green gemstones, peridot, green sapphire, chrysoberyl, andalusite, enstatite and, perhaps one should say, emerald, may somewhat resemble demantoid; but each of these is doubly refracting between crossed polaroids, gives refractometer readings, and has a distinctive absorption spectrum of its own. The spectrum of emerald and of the chrome-bearing chrysoberyl, alexandrite, have already been fully described, and the others will all be dealt with in due course. Only the yellow-green chrysoberyl has a spectrum at all like that of demantoid, in that its main feature is a single band in the beginning of the violet. In chrysoberyl, however, the band is broader and more diffuse, and never attains the savage intensity of the demantoid band.

Unorthodox test

Demantoid is, indeed, one of the easiest of stones to recognize at sight and to prove by simple tests. As an instance of this one may recollect an actual routine test which had to be carried out on a parcel of 117 so-called 'olivines' by one of us during the war[1]. In this instance it was decided that a rapid run through under the microscope would be the best preliminary manoeuvre – and so it proved, since 110 of the stones revealed themselves as undoubted demantoids by virtue of their typical inclusions. The seven stones remaining were then checked by means of the spectroscope. One was a 'clean' demantoid, one was a green sapphire, showing the typical threefold group of bands in the blue; two were peridots, with three widely spaced bands in the blue; and three yielded an exquisitely sharp rare-earth spectrum which was recognized as belonging to a certain type of green andalusite from Brazil. No orthodox tests were therefore needed for this parcel, and the whole very pleasant proceedings occupied less than an hour. The spectroscope could, as it happened, have been used throughout, but would have been rather more strain on the eyes for so many stones which, being small, would have needed careful positioning to give clear results.

18 Other Chromium Spectra

The absorption spectra of all those important gem varieties which owe their colour to chromium have now been described. There remain a few rare examples of chromium spectra to be considered before passing on to stones which show absorption bands due to iron. These rare stones are only interesting to the specialist, and can be dealt with quite briefly. In most of them the spectra are elusive and some may be seen only in the finest stones.

Hiddenite $Li_2O.Al_2O_3.SiO_2$

Spodumene of the almost emerald-green tint to which the name hiddenite was originally given in 1881 is probably only found in North Carolina, USA; though the name has since been used also for yellowish-green spodumene from Madagascar and Brazil. All these varieties show a strong narrow absorption band due to iron at 437.5nm, analogous to that shown by jadeite; but the green specimens from North Carolina have in addition a fairly rich assemblage of chromium lines in the red, which give a clue to the origin of its attractive colour. A beautiful example of this true hiddenite can be seen in the Mineral Gallery of the British Museum (Natural History), where a faceted stone of a lovely green tint, weighing 2½ carats, is displayed.

The usual strong chromium doublet in the deep red is seen in hiddenite at 690.5 and 686nm, and there are weaker lines at 669 and 646nm. There is a broad absorption band centred near 620nm, but no chromium lines are to be seen in the blue. The strong and well-defined iron band at 437.5nm already mentioned is accompanied by a weaker band, also fairly narrow, at 433nm. In yellow spodumene from Burma a narrow line at 505nm has also been observed.

Euclase Be(AlOH)SiO₄

This rare beryllium mineral is much prized by collectors, both in the cut and uncut state. To the lapidary it is something of a nightmare owing to its perfect and distressingly easy pinacoidal cleavage. The colour of euclase usually resembles that of aquamarine rather than of emerald, but in the deeper specimens the distinctive lines of chromium can be clearly seen. The customary strong doublet is of unusually long wavelength, 706.5 and 704nm being our measurements for this. Much weaker lines were measured at 695, 688, and 660nm, while broader bands were seen at 650 and 639nm. Finally narrow but ill-defined bands were also measured in the blue at 468 and 455nm. This spectrum was seen in an exceptionally fine cut stone belonging to Mr R.C. Mathews, and might be difficult to see in a paler euclase.

Kyanite Al₂O₃.SiO₂

With its blue-and-silver, bladed crystals, kyanite is one of the most beautiful minerals known to the collector. Its flaky, easily cleavable nature, however, make it hardly suitable for cutting as a gemstone. Though normally it is blue – almost a sapphire blue – as its name suggests, it can also be green or almost colourless. Its absorption spectrum is very unreliable, but blue specimens from a chrome-rich locality in Rhodesia showed a distinct chromium spectrum. A very strong line (possibly a close doublet) was measured at 706nm, another strong line at 689 while there were weaker lines at 671 and 652nm.

Dark blue kyanite from another source showed bands of moderate strength in the blue at 446 and 433nm, and a photograph of a spectrum of this stone revealed further bands in the near ultra-violet at 370, 357 and 340nm. These last five bands were not due to chromium but probably to iron. Many kyanites are disappointing to the spectroscopist in showing no bands at all.

With the exception of demantoid, all the minerals showing chromium spectra which we have so far described contain *aluminium* as an essential constituent, and the chromium has in each case replaced a small percentage (usually between 0.1 and 2.0%) of the aluminium in the crystal lattice. The presence of a clear-cut chromium spectrum is, in fact, strong presumptive evidence that one is dealing with an aluminium mineral. There are admittedly a few gemstones which contain no aluminium or other trivalent element and yet may show a chromium spectrum, though this is never as sharply-defined as in minerals containing alumina. Enstatite and diopside, for instance, quite often contain enough chromium to influence the colour favourably and show lines in the red, and even peridot from one locality (Hawaii) contains its small quota of chromium. Woolly chromium bands can also be discerned in chalcedony which has been stained with chromium salts to resemble chrysoprase, and certain green glasses betray the presence of chromium by faint banding in the red.

We will now describe these briefly in turn.

Enstatite

This has a striking and prominent narrow band in the green at 506nm, but this and other bands in the green and blue are due to iron, and will not be given detailed treatment until later. The chromium bands in the red which concern us at the moment are to be seen notably in the pretty green pebbles from the Kimberley diamond fields which are found in company with pyrope garnet. These are seldom large enough to yield cut stones of more than a carat, but are prized by discriminating collectors. The two lines measured lie near 688 and 669nm, and each has a transparency-patch alongside, giving the peculiar appearance already noted in some of the lines in emerald.

Diopside

Diopside, or perhaps one should specify *chrome diopside*, which, like enstatite, belongs to the pyroxene family, has a very similar spectrum. There are, however, two narrow bands at 508 and 505nm in place of the one in enstatite, and the chromium lines are stronger, and situated at 685, 657 and 634nm.

Peridot

Attractive pebbles of *peridot* from the beaches of Hawaii were sent to us by Miss Gwynne Richards of New York. In addition to the usual iron spectrum in the blue we were very surprised to note undoubted chromium lines in the red at 687, 667 and 635nm. Later, we found from the literature[1] that chromium had been estimated quantitatively (0.14%) in a chemical analysis of peridot from this source. Hawaii peridot is also notable for the bubble-like nature of its numerous inclusions.

Green stained chalcedony

Usually recognized by its rather unpleasing blue-green colour. The presence of vague chromium lines centred at about 705, 670 and 645nm serves to discriminate between this material and true chrysoprase which owes its colour to nickel and shows no absorption bands. An occasional *green glass* has been encountered in which ill-defined chromium bands could be seen.

Three final examples of chromium spectra may be mentioned. *Green aventurine quartz* owes its colour to flakes of the green chrome-bearing variety of muscovite mica known as fuchsite, which can also be seen as an inclusion in emeralds from Siberia and from the Transvaal. The green aventurine shows bands at 682 and 649nm approximately, which are undoubtedly due to the mica.

It was Robert Webster who first drew our attention to narrow absorption bands in the translucent green mineral *variscite*. From their position and nature it seems probable that these are due to chromium. Bands were measured at 688 (strong) and 650nm (weak and vague).

Violet-blue *scapolite* from Burma shows strong absorption of the yellow and may also have narrow bands at 663 and 652nm, which seems to indicate a chromium absorption. This material is sometimes faceted, but is perhaps more often cut *en cabochon* to bring out a cat's-eye effect. It is strongly dichroic and the bands belong to the deep-coloured ordinary ray.

19 Absorption Spectrum of Almandine Garnet

It is much less easy to describe in general terms the features which absorption spectra due to iron have in common than it was in the case of chromium. In the first place, iron spectra ought to be considered in two separate categories: those in which the bands are due to divalent or 'ferrous' iron, corresponding with ferrous oxide (FeO), and those in which they are due to trivalent 'ferric' iron, corresponding with ferric oxide (Fe_2O_3). Almandine, peridot and blue spinel provide typical examples of ferrous iron spectra; while chrysoberyl, yellow orthoclase and green sapphire are among the gemstones which show bands due to ferric iron.

Ferrous iron is often found partially replacing magnesium, which is also divalent and has a very similar ionic radius (Fe'' 0.083; Mg'' 0.078nm) and although in almandine, for instance, ferrous iron is the predominant partner, magnesium is always present in significant amounts. From the examples given above, it will be seen that divalent iron can give rise to red, green or blue minerals. The colours can be quite attractive, but are certainly not so bright or clear as those produced by chromium. This is due to an overall absorption of the spectrum colours, which is not confined to the visible bands. Absorption maxima are commonly to be found in the blue and the green regions of the spectrum, and the bands have a tendency to be broad and diffuse, though often a fairly dense and narrow 'core' may be discerned within some of the broad bands, and by no means necessarily in the centre of the diffuse absorption region. That is to say, if an absorption curve is drawn, the peak representing the absorption band will not be symmetrical. In measuring the wavelength of such a band, our practice is to give the measurement for the summit of the peak rather than the arithmetic mean of the limiting wavelengths of the entire band.

We can now give details of individual spectra due to ferrous iron, starting with that of almandine garnet.

Almandine garnet, $3FeO.Al_2O_3.3SiO_2$., shares with zircon the distinction of being mentioned in the short but important note which Sir Arthur Church sent to the *Intellectual Observer* in 1866[1]. This gave the first indication that gemstones might readily be identified by means of their absorption spectra. After discussing the effects seen in zircons of different kinds, Church writes: 'The iron-garnet of different shades (carbuncle, almondine [sic], etc.) gives a beautiful and very characteristic spectrum with several intensely deep absorption bands.' Mr A.E. Farn was fortunate, some years ago, to come across a volume of the *Intellectual Observer* containing Church's letter, on a second-hand bookstall. In a short article[2] recounting his discovery, the letter from Church was reprinted in its entirety.

It is hardly conceivable that Church, having made so exciting an observation in the case of almandine and zircon and being an enthusiastic collector and student of gems, did not examine other coloured stones with his spectroscope in the hope of seeing bands in some of these also. Since successive editions of his book *Precious Stones* (which he last revised in 1913, when he was seventy-eight years old) continue to mention only zircon and almandine in this connection, he must have failed in any further attempts he made – perhaps through use of an unsuitable spectroscope. Be that as it may, the legend persisted for more than sixty years after the initial discovery that in only the two gemstones mentioned were characteristic spectrum bands to be seen.

Short of holding the instrument the wrong way round, no observer with a spectroscope could fail to see the absorption bands in almandine garnet: the spectrum is easier to see and to identify than that of any other gemstone. The three main absorption bands are both broad and strong, and are situated near the centre of the spectrum, where the visual acuity is good.

The first of these is in the yellow at 576nm and is about 20nm broad. The other two are in the green, centred at 526 and 505nm, so that the middle band is perceptibly nearer to the third than to the first band, even when viewed through a prism spectroscope in which the dispersion widens as it approaches the violet end. 576 is a strong band, but 505 is most certainly the strongest of the three, while 526 is weaker and ill-defined, often merging with 505 to make one broad band in dark red stones. This can be verified by observing the variation in intensity between the absorption spectra of pale and dark specimens.

It is worthwhile to note that red and orange wavelengths above the 576 band are always brightly transmitted and the gap between 576 and 526 is again fairly bright, but that between 526 and 505 is usually gloomy with absorption, as is the whole blue end of the spectrum.

The diagnostic importance of the 505 band, and the comparative sharpness of the 'core' which is found within it, led us to attempt a more accurate measurement of its position than is possible with the Beck 'Wavelength' spectroscope which we normally employed. Using a table spectrometer in conjunction with a diffraction grating, comparisons of the almandine band were made against the accurately-known emission lines of lithium (497.2), barium (493.4) and cadmium (508.6nm).

This experiment proved the centre of the 'core' of the chief almandine band to have a wavelength of 505.2 ± 0.3nm.

Under good conditions, almandine shows several other bands in addition to the three prominent ones already described: one in the orange, at 617 and another in the blue at about 462nm being particularly noticeable (see Fig. 19.1). Below is a list of bands measured by us in a specimen having a refractive index of 1.77, and density 3.95. The spectrum was also photographed.

617nm	m.w.	broad
576	s.	broad
526	m.s.	broad
505	v.s.	broad (with core)
476	w.	narrow
462	m.	narrow
438	w.	narrow
428	m.w.	broad
404	m.w.	narrow
393	m.w.	narrow

The bands at 404 and 393nm, being in the extreme violet, are not discernible in the general gloom at the end of the spectrum, but they were clearly visible on the photograph. The 462 band is in almost exactly the same position as one of the bands found in spessartine garnet, and since three per cent or more of this manganese garnet are to be found in many almandines the band might be thought due to manganese rather than to iron. But a far stronger spessartine band is one at 432nm, and this is missing from almandines which show the 462 quite clearly. The latter therefore can be confidently credited to the iron garnet.

Absorption of Gem Minerals

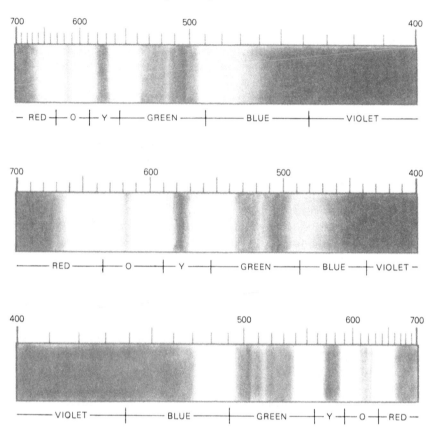

Fig. 19.1 ALMANDINE GARNET. Three intense bands due to ferrous iron provide an instantly recognizable spectrum, traces of which can be seen in other red garnets

It can be said that almost all red garnets show the almandine spectrum to a greater or lesser degree. In stones containing the almandine molecule 'diluted' with pyrope (magnesium) and grossular (calcium) which have no bands of their own, the proportion of almandine present can be estimated fairly closely by the strength of the absorption bands seen – though refractive index is admittedly a surer guide. In the chrome-rich pyropes described in a previous chapter, the appearance of the spectrum is altered, since a broad chromium absorption region in the green masks two of the three almandine bands present, leaving only the 505 'core' showing, and that not very strongly. Specimens of the rare and attractive manganese garnet, spessartite, often contain some almandine and will then show the three main almandine bands in the green in addition to their increasingly strong series of manganese bands in the violet, which will be described in detail later. Hessonite, which has no spectrum of its own, may also show a faint almandine spectrum, and traces of manganese bands as well.

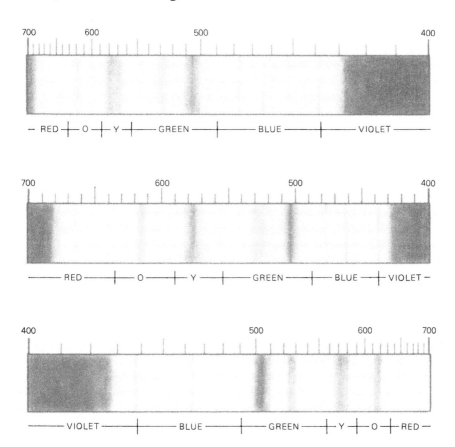

Fig. 19.2 LIGHT PYROPE/ALMANDINE. A paler version of the almandine spectrum often seen in combination with other garnet spectra

Finally it should be remembered that the almandine spectrum may be noticed in the garnet-topped doublets often found in antique jewellery, and which can deceive the unwary. The garnet used is invariably an almandine; but since only a thin slice is present and the coloured glass base may contribute (if blue) cobalt bands of its own, or (if red) a broad band in the green, only the 505 band of the almandine may be detected unless the stone

131

can be tilted on edge and light which has passed through the garnet top only can be analysed.

The almandine spectrum is unlike the absorption spectrum of any other red stone, so that it is one of the most useful diagnostic spectra available to the gemmologist. Garnets in old jewellery, for instance, often with foiled backs and curved surfaces, can be quickly checked with the spectroscope by reflected light, and garnets which show no reading on the refractometer on account of their high index cause no trouble when a spectroscope is handy. The broad cobalt bands seen in synthetic blue spinel and in many blue glasses may be said to have a certain superficial similarity to the three main bands of almandine, though their position and distribution are very different: but the body-colour of the specimen itself here prevents any possibility of confusion.

Figs. 19.1 and 19.2 depict respectively the strong spectrum of a deep coloured almandine and the less emphatic one seen in a typical lighter 'pyrandine' garnet.

20 Absorption Spectrum of Blue Spinel

The absorption spectrum of blue spinel is neither so spectacular nor so easily recognized as that of almandine garnet and though both are undoubtedly due to ferrous iron it is difficult to trace much connection between them. To the practised eye the spectrum of blue spinel is none the less very distinctive, and its most salient features can be traced also in purple or almandine-coloured spinels and in the paler, mauve, types which are not so easily identified at sight. When dealing with chromium spectra we have shown how the fluorescence spectrum of red and pink spinels can be used as a sure and sensitive means of identification. It is thus fortunate that the presence of iron, which 'kills' this fluorescence effect, brings with it absorption bands of its own which are almost equally helpful.

The only published reference to the blue spinel spectrum which we have seen is to be found in the paper by E.T. Wherry[1] in which he mentions two strong broad bands centred at 550 and 460nm, with weaker bands at 590 and 510nm, ascribing them to the presence of cobalt. There is admittedly a certain resemblance to a cobalt spectrum in the arrangement of the bands in the orange, yellow, and green seen in the natural blue spinel; but in this instance we are in the strong position of being able to make a direct comparison with the bands induced by cobalt in the spinel lattice by studying synthetic blue spinels made by the Verneuil process, in which cobalt is definitely present as a colouring agent. It can then be seen that both in their nature and position the bands in the natural spinel differ from those in the cobalt synthetics. Moreover the strongest bands in the natural mineral are in the blue region, where cobalt has no bands to show.

In describing the spinel spectrum it will be well to mention first these bands in the blue, since they are the most powerful and persistent of the series. There is a broad band in the deep blue centred at 458nm and a much narrower band of about equal strength at 478nm, separated from the other by a distinct gap. The drawings (Fig. 20.1) will give a better idea than words can of the disposition of these and other bands in the spectrum. If the 458 band were unaccompanied it might be confused with the group of three bands seen in some sapphires, though the complex nature of the latter is at variance with the homogeneous block of shadow seen in spinel. But the narrow 478 band alongside makes a very distinctive pattern which is easily recognized. In addition to these strong bands in the blue there are a number of others in the green, yellow and orange which are more difficult to 'pin down' exactly in description or measurement but which, taken together, present an appearance which is unique amongst gemstones, and at its best

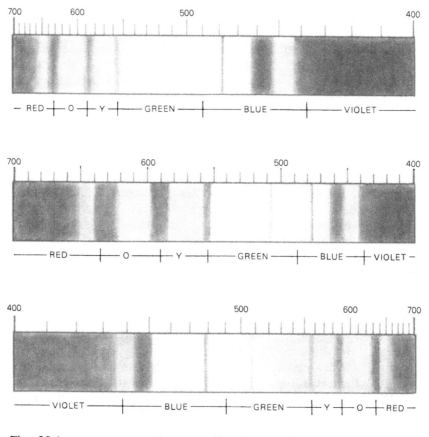

Fig. 20.1 BLUE SPINEL. A generally gloomy looking spectrum of vaguely defined bands

can be very beautiful. What appear at first glance to be broad vague bands can be resolved in good specimens into doublets – or perhaps it would be more exact to say that within these vague absorption regions two distinct nodes or 'cores' can be detected.

In the following list of our measurements the wavelengths of these 'bands-within-bands' are given, together with the mean position of the broad bands taken as a whole.

$\frac{640}{624}$ > 635nm m. broad

$\frac{592}{577}$ > 585nm m. broad

$\frac{558}{553}$ > 555nm m. broad

508nm w.	narrow
478nm s.	narrow
458nm s.	broad
443nm w.	narrow
433nm w.	narrow

The last two bands in the violet are seldom seen, but are included for completeness.

The only stone with absorption bands which in their 'pattern' resemble at all closely those in blue spinel is that rare gem, taaffeite. Only two specimens of taaffeite have been examined by us. Since both of these were small and pale their absorption spectra were exceedingly weak. But what bands could be seen resembled those in spinel to a marked degree in their nature and distribution. This indicates not only a common cause (ferrous iron) for the phenomena but also a similarity in the internal structure of the crystal. Now taaffeite, though possessing full hexagonal symmetry, might well be considered from the chemical point of view to be a spinel (using the term spinel in its group significance). It is a beryllium magnesium aluminate, and is thus intermediate in composition between magnesium aluminate (spinel) and beryllium aluminate (chrysoberyl). Despite the difference in symmetry, which is due to the small atomic radius of beryllium, the structure of taaffeite is undoubtedly very close to that of spinel.

Of the other blue stones which might be mistaken for spinel, the blue or blue-green form of tourmaline known as indicolite is the closest in colour and general appearance. The spectrum of indicolite consists usually of a single rather narrow band at 498nm

135

– just where the green region merges into the blue, and a broader, vaguer band near 460. On paper, these two bands – a broad band in the blue with a narrower one on the green side of it – might easily be confused with the blue bands in spinel. Actually, however, their appearance as well as their position is distinctly different, and indicolite further lacks any bands in the green and yellow regions which make such an effect in the blue spinel spectrum.

Those sapphires which have the rather inky or greenish blue of spinel are usually Australian, containing more iron than those from other localities. This means that they have a fairly strong version of the spectrum which is seen at full strength in green sapphire: a strong band at 450, merging into another at 460 and a fainter member at 471nm. Even though these three bands are strong enough to merge into a single block this is clearly a complex one, and should not be confused with the homogeneous 458 band of blue spinel which, moreover, has the narrow band at 478 also present to help the observer in his identification. Some deep-coloured sapphires may show a distinct, broad absorption band in the yellow-green which might be mistaken by a beginner for the far more elaborate series of bands in spinel. If there is any doubt, a check with the polariscope or dichroscope will put matters on a sure basis, and so, of course, will a refractometer reading.

It may be remembered that most of these Sri Lankan blue or bluish spinels have traces of zinc, and some contain much more than traces (up to 18 per cent was found in an earlier research on the subject). These 'gahnospinels' as they are called (a term compounded from gahnite, the zinc spinel proper, and spinel) have markedly higher density (up to 3.98) and refractive index (up to 1.748) than normal, but there is no clue in their colour or appearance as to whether they are zinc-rich or not. Thus, as one might expect, the spectroscope gives the same range of bands with gahnospinels as with any other blue spinel of equal tint.

Synthetic blue spinel

Confusion does not often arise between natural and synthetic blue spinel, because the bright cobalt-blue colour of the Verneuil product is so markedly different from the rather sad and restrained colour of the mineral. Most students, when in doubt, use a Chelsea colour filter, under which the red or chestnut colour of the cobalt synthetic is very striking. It should be remembered, however, that natural blue spinels can also appear distinctly reddish under the filter, though certainly not the spectacular red of the blue synthetic.

With the spectroscope, the three characteristic and homogeneous cobalt bands are easily recognizable in the synthetic blue spinel, varying in strength with the depth of colour. With a prism spectroscope, the centre band is notably wider than the others, which serves to distinguish the stone from cobalt-blue glass which otherwise has a very similar spectrum.

Cobalt in natural spinel

In the early 1970s four small spinels of an exceptionally fine gentian-blue were purchased from two different sources in Sri Lanka. Their colour was quite different from that of the normal blue natural spinel and very reminiscent of synthetic cobalt-blue spinels. Like these, the stones were a strong red when viewed through the emerald filter, and at first sight were dismissed as synthetic, especially when their spectra showed what appeared to be a cobalt absorption. But further examination gave R.I. of 1.717 and revealed natural crystal inclusions. These stones were eventually written up in 1977[2].

In 1984 J.E. Shigley and C.M. Stockton published a detailed paper[3] covering their researches on some eighteen of these exceptionally blue spinels, comparing them with ten known synthetics. Up to 0.05 per cent of cobalt was found in the natural stones and the cobalt absorption spectra were identified with certainty. However, all these natural cobalt spinels showed an iron band at about 460nm which was not present in any cobalt synthetic, so they were still capable of being separated from one another.

21 Absorption Spectra of Peridot and Sinhalite

We have already described the absorption spectra of two gemstones owing their colour to ferrous iron. One of these (almandine) is *red*, the other (spinel) is *blue*, while peridot and sinhalite, the subjects of this chapter, are *green* and *yellow* or *brown* respectively. Thus we see that iron is even more versatile than chromium in the range of colours it can induce, even if the colours themselves do not show the same richness and intensity.

Peridot is the jeweller's name for a green intermediate member of the important olivine family of minerals, which form an orthorhombic isomorphous series ranging from the magnesium silicate forsterite at one extreme to the iron silicate fayalite at the other, with manganese sometimes playing a part. The name peridot is unequivocal, and therefore to be preferred to the English mineralogist's 'olivine' and to the American mineralogist's 'chrysolite', both of which have so often been used in other connotations that they give rise to confusion.

Neither forsterite nor fayalite have been used as gems, and it seems rather curious that there is so marked a similarity in colour, physical properties, and thus, presumably, in composition between gem-quality peridot from all the known localities – Zeberged, Burma, Arizona, Congo, Norway, Hawaii, etc. These have a position in the series corresponding to about 10 per cent of FeO. The density of all true peridots which we have tested falls within the narrow range 3.33-3.35, and the beta refractive index varies only between 1.666 and 1.673, the birefringence being 0.036 or a little lower. The brown or brownish peridots sometimes encountered have very similar properties to the above and are quite easily distinguished from sinhalite, which was formerly accepted as an iron-rich form of peridot.

The absorption spectrum of peridot is distinctive and its main features are soon described. There are three main bands in the blue, well separated and evenly-spaced, centred at 493, 473 and 453nm. The 493 band has a distinct narrow 'core' at 497nm; the 473 band is also fairly narrow, while the 453nm band is broader and less well-defined. In large specimens, weak, vague bands in the orange (635) and green (529) can be discerned. The only other bands we have measured are two situated in the near ultra-violet, which were clearly recorded on a photograph of the peridot spectrum; these are centred at 397 and 385nm, approximately.

As with most anisotropic stones, there is a considerable variation in the nature and strength of the absorption bands in accordance with the vibration-direction of the transmitted light. Something of this variation can be noted simply by altering the orientation of the stone; but for a clear-cut separation of the three spectra belonging respectively to the alpha, beta, and gamma vibration directions it is necessary to use a polarizer, and either a crystal of peridot or a cut stone in which the orientation is known.

The following measurements were made by using these methods:

α	495nm	m.s.	
	473nm	v.w.	
	453nm	v.w.	
β	529nm	w.	vague
	497nm	v.s.	complex
	473nm	m.s.	narrow
	453nm	m.s.	broader
γ	493nm	m.s.	narrow
	473nm	m.w.	narrow
	453nm	v.w.	broad

When using the spectroscope for identification purposes it is quite unnecessary to go into such detail. But in cases where the spectrum is so weak as to be difficult to distinguish it may be of practical help to interpose a polaroid film either below or above the spectroscope and turn this until the bands appear to be at their strongest.

The spectrum of sinhalite

We have already said that the rare mineral taaffeite has a very similar spectrum to that of blue spinel. By a most curious coincidence, the next new gem mineral to be discovered (named sinhalite[1] by Dr G.F. Claringbull, who was the first to establish it as a new species) has an absorption spectrum showing an equally close resemblance to that of peridot. Despite the fact that sinhalite is a magnesium aluminium *borate* (containing some iron) while peridot is a magnesium-iron *silicate*, the crystal structures of the two minerals are very similar, as proved by X-ray measurements on the cell dimensions. Thus, once again, we see the way in which a strong similarity in the absorption spectra of two minerals points surely to a similarity in structure even where they belong to a different crystal system (spinel, taaffeite) or have an entirely different composition (peridot, sinhalite).

One final coincidence in the way these two new gem materials 'tie-up' with each other is that peridot and spinel are to some extent structurally related, as pointed out by J. de Lapparent. In spinel one finds tetrahedra of MgO_4 linked by octahedra of $Al(O_6)$; while in peridot one finds tetrahedra of SiO_4 linked by octahedra of $Mg(O_6)$.

Unlike taaffeite, sinhalite is moderately common, though so far Sri Lanka seems to be the only gem source of the mineral. Years before its true nature was guessed at it was cut by the native lapidaries and mixed in with parcels of zircons, chrysoberyls, or tourmalines of similar colour, from which by eye it is difficult to distinguish. Those of us who measured its constants regarded it as an interesting iron-rich variety of peridot, since its specific gravity (3.46-3.52) refractive indices (beta ranging from 1.697 to 1.704) and birefringence (0.037-0.038) accorded very plausibly with what one would expect from a member of the olivine family. Had the absorption spectrum of sinhalite been markedly different from that of peridot we should have taken further steps to confirm its identity, but in actual fact it is so very similar in most of its aspects that our suspicions were lulled.

Absorption bands in sinhalite are as follows:

526nm	w.	vague
493nm	m.s.	narrow
475nm	m.s.	narrow
463nm	m.s.	broader
452nm	m.	narrow
435nm	s.	as cut-off

Comparison with the wavelengths for peridot given earlier will show that the three main bands in the blue are virtually duplicated by the bands in sinhalite. The only really significant difference is the presence of an extra band in sinhalite at 463nm. The band sometimes seen in peridot at 635nm has not been noted in sinhalite, but this is in any case so weak and vague as to be of small diagnostic value. A careful study of the absorption bands in the two minerals, however, does reveal certain differences in detail which assists in distinguishing them. The sinhalite bands do not show the complexity evident in some of the peridot bands, and they are not so strongly influenced by variations in vibration direction. The 463nm band in sinhalite is noticeably broader than the others and is flanked by two narrow bands, while in peridot it is the end band at 453nm which is seen to be the broadest of the group. Another difference (though this is chiefly of academic interest) is that the bands in the near ultra-violet, which we have recorded photographically in peridot, have not been seen on photographs of sinhalite spectra.

Summary

The beautiful olive-green colour of peridot is not matched by any other gemstone, so that a stone of this particular colour is almost certainly a peridot or a paste. Two simple tests confirm this: observation of 'doubled' facet edges with a lens, and a glance at its absorption-spectrum, which is characterized by three rather narrow and widely spaced bands in the blue. If the specimen is small or pale in colour these bands may be hard to see. Viewing the stone sideways or on end may strengthen the effect in such cases, or a piece of polaroid may be interposed and turned until the optimum strength of the bands is found. But some large deep-coloured peridots absorb strongly in the violet and blue and there is an apparent cut-off at the 497nm edge of the first band in the blue.

Sinhalite is commonly pale to deep yellow or brown in colour, though certain specimens if suitably orientated by the lapidary may show a distinctly greenish cast. Its appearance resembles chrysoberyl and zircon most closely, and from these it may easily be distinguished by its absorption spectrum and of course by many other properties. A much more difficult task is to separate sinhalite from the (very rare) brown peridot. So far as the spectrum goes, the answer lies not in careful measurement but in observation of the last three bands in the blue. In sinhalite these consist of a broad band flanked by two narrower bands, and a fourth band will be seen on the border of the green and blue. In peridot there are only three bands in all for the blue region, and the middle band here is noticeably narrower than the end band which is broad and vague. In certain cases it may not be possible to see the spectrum clearly enough to enable one to make a definite decision. In such cases one must resort to more orthodox methods. As shown by the figures given above, there is a distinct gap between the highest peridot density and refractive index and the lowest sinhalite figures for these constants. On this basis, therefore, a distinction between the two minerals can be made with confidence.

The drawings (Figs. 21.1 and 21.2) will enable the reader to visualize the slight differences between the absorption bands in peridot and sinhalite which are described in the text.

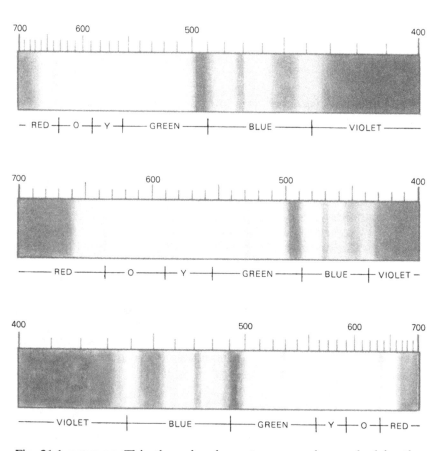

Fig. 21.1 PERIDOT. This three band spectrum may be masked by the general absorption of the blue-violet and may be seen as a total cut-off at about 500nm. Stronger lighting may allow the full structure to be seen

Peridot and Sinhalite

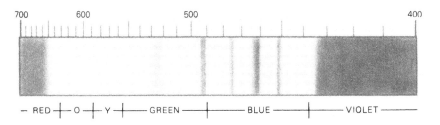

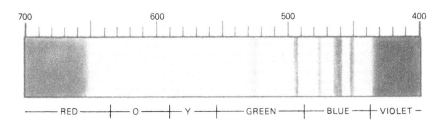

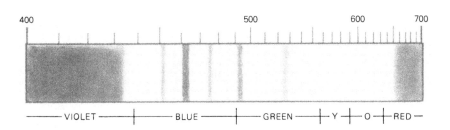

Fig. 21.2 SINHALITE. The three bands seen in peridot are here again at almost the same wavelengths, but the extra band at 463nm identifies sinhalite, which is in any case a yellow or brown gem rather than peridot green

143

22　Absorption Spectra of Enstatite and Diopside

The last chapter was concerned chiefly with the absorption characteristics of peridot – the only member of the important olivine family of minerals which is utilized as gemstone. We now turn to two members of another great group of rock-forming minerals, the pyroxenes, which includes such well-known gem materials as jadeite, spodumene and rhodonite. Since at present we are confining our attention to stones which owe their absorption spectra to ferrous iron, we will deal now with two pyroxenes, enstatite and diopside, which are cut more to please collectors than for use in commercial jewellery.

The first of these, enstatite, is an orthorhombic mineral. Just as peridot is one of a series of magnesium and iron *orthosilicates*, $(Mg, Fe)_2 SiO_4$, so enstatite belongs to a series of magnesium and iron *metasilicates* $(Mg, Fe) SiO_3$. The name enstatite should properly be reserved for specimens containing up to about 5% of iron: hypersthene is the term for iron-rich members of the series (over 15% iron), while bronzite can be used for intermediate examples although when the mineral shows no schiller effect this name seems hardly appropriate.

Diopside, the second mineral whose spectrum we shall describe, is a monoclinic pyroxene which is essentially a silicate of calcium and magnesium, $Ca,Mg(SiO_3)_2$, in which ferrous iron replaces some of the magnesium. As we have learned to expect from minerals which are structurally related and contain the same colouring agent, there is often a close similarity between the spectra of these two minerals.

The spectrum of enstatite

The absorption spectrum of enstatite is so striking that we were surprised to find no mention of it in Wherry's comprehensive

144

paper, nor elsewhere in the literature. The salient feature is a single sharp line in the blue-green at 506nm (an accurate measurement on the table spectrometer gave 506.2nm) which is moderately strong in the attractive green enstatite pebbles which accompany diamond and pyrope in the Kimberley district of South Africa and is positively intense in the more iron-rich brown enstatites or 'bronzites' from Mysore and from Burma, which have sometimes been referred to as hypersthene-enstatites.[1,2]

Almost colourless crystals of enstatite from Sri Lanka, examined in 1985, were found still to show the sharp 506 line and a vague band centred at 548nm. An iron spectrum in an almost colourless gem is quite extraordinary. To the enthusiast a sharp and intense absorption band gives a peculiar thrill of pleasure, and the 506 line in this mineral is certainly one of the most striking to be seen in any stone. Notable lines in other minerals, such as the diamond 415.5nm and even the zircon 653.5nm, are situated near the visible limits of the spectrum and therefore may not easily be seen by all observers, whereas the enstatite band is well towards the middle of the spectrum and can hardly be missed by anyone, no matter how inexperienced.

There are several other enstatite bands to be seen in favourable specimens, though these are relatively weak and vague. The chief of these is a broadish band in the green at 548nm, while there is a series in the blue more difficult to see and to measure, a strong band forming the end of the spectrum in the violet, and two faint narrow lines flanking the famous 506 position. Here are these bands in the form of a table:

548nm	m.	broad
509nm	v.w.	narrow
506nm	v.s.	narrow
502.5nm	v.w.	narrow
483nm	m.w	broad
472nm	m.w.	broad
459nm	m.w.	broad
449nm	m.w.	broad
425nm	s.	broad

It is interesting, and perhaps significant, that ferrous iron spectra often show an important band at or near 505nm. The most notable case is almandine, with its strongest band at this wavelength. Blue spinel shows a (weak) band at 508nm, and kornerupine absorbs at 503nm. In peridot and sinhalite, admittedly, the nearest corresponding band is some 10nm lower, but one must remember that variations of ten or twenty nanometers are to be found in the position of the deep red chromium doublet in different species (e.g. euclase, 706, 704nm; emerald, 683, 680nm) and in these there can be no doubt at all that the lines have a common physical cause and only vary in wavelength on account of the influence exerted by the surrounding atoms in the crystal lattices concerned.

Except for certain types of diopside there is only one other mineral which shows so narrow and sharp an absorption line near 506nm, and strange to state that mineral is diamond.

(At this point in the original paper (*The Gemmologist*, Vol,24, p.103) Anderson went on to show that the heat-sensitive 504 line in natural and treated green and brown diamonds could not be due to iron. We now know that this line in diamond is due to included nitrogen.)

Absorption in diopside

Diopsides of gem quality vary in colour from very pale to a dark and rather sombre green. The most pleasing greens are seen in 'chrome-diopside' types, and it is only in these that a striking spectrum is seen which is so similar to that shown by enstatite. Our measurements were taken from some attractive chrome-diopside catseyes from the Mogok stone tract in Burma. In sunlight or under a single lamp these show a curious 'forked' chatoyant ray. This can hardly be called a star, since the angle between the 'prongs' is only about 12°. With these stones, in place of the single prominent line in the blue-green seen in enstatite at 506nm, there are two lines of almost equal strength at 508 and 505nm. There is a broad band in the green almost exactly similar to that in enstatite – but only one band in the blue (490nm) has been measured. Since these stones contain chromium there are also typical chromium lines in the red: a strong doublet at 690nm accompanied by three rather woolly lines towards the orange. For convenience, all these lines and bands are tabulated below.

690nm	S.	doublet
670nm	m.	woolly
655nm	m.	woolly
635nm	m.	woolly
547nm	m.	vague
508nm	m.s.	narrow

505nm   m.s.   narrow
490nm   m.   vague

In the duller green diopsides the spectrum is not nearly so attractive or diagnostic. In a leaf-green stone from Madagascar, for instance, rather ill-defined bands were measured at 505, 493 and 456nm.

The drawings (Figs 22.1 and 22.2) show the similarities and differences between the enstatite and chrome-diopside spectra very well. The chromium bands seen in Kimberley enstatites are here added to the strong spectrum typical of a brown iron-rich enstatite.

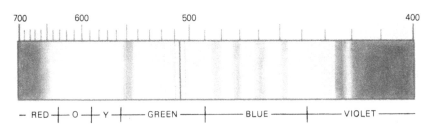

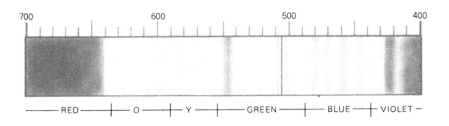

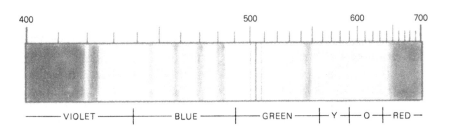

Fig. 22.1 ENSTATITE. Another startlingly recognizable spectrum with an intense line at 506nm, supported in brown enstatite by a faint line on either side

Absorption of Gem Minerals

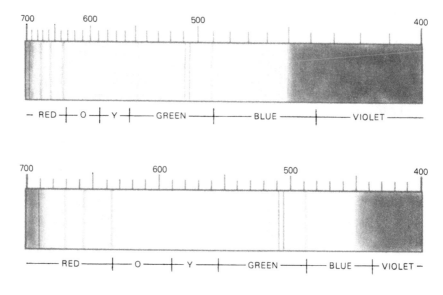

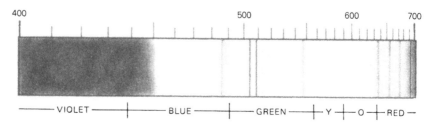

Fig. 22.2 DIOPSIDE. The green chromium type may have lines in the red but the most prominent absorption is a pair of lines either side of the position of the enstatite single line

Summary

Enstatite has an exceedingly distinctive spectrum: all specimens so far examined showing a clear-cut narrow line at 506nm in the blue-green. This is at its strongest in brown enstatites, and other bands can be seen in such specimens – notably a broadish band in the green. Amongst those stones which might be confused with enstatite on grounds of appearance only chrome diopside can show a spectrum at all resembling it. The most easily checked difference is that the diopside shows a double line in place of the single line in the blue-green of enstatite. Another difference between the two is that when a polaroid disc is rotated in the optical path while viewing these spectra, very little difference in the strength of the bands can be noted in the case of enstatite, whereas in chrome diopside dichroism is evident in the considerable variation in the strength of the bands when the disc is turned. The ordinary dull green forms of diopside show only a few feeble bands which have small diagnostic value.

23 Absorption Spectra of Kornerupine and Axinite

Before the account of absorption spectra due to ferrous iron can be completed there are still several minerals to be discussed which are occasionally cut as gems, though seldom seen in commercial jewellery. We now deal with two of these: kornerupine and axinite.

Kornerupine

Kornerupine is a complex borosilicate of magnesium and aluminium containing variable amounts of iron which, according to Dr M.H. Hey, is present mainly in the ferrous form. The mineral was named in memory of A.N. Kornerup by his friend J.T. Lorenzen, who first investigated the mineral when it was discovered at Fiskernaes on the west coast of Greenland.

Kornerupine is one of the rarer minerals which have been found among the gem gravels of Sri Lanka; until recently the chief source of the gem quality material. From this area it is usually a dark brown-green in colour with little to distinguish it from the many dark tourmalines which abound in these gravels, so the fact that it occurred at all in Sri Lanka was not realized until late in the 1930s. That other great storehouse of gems, the Mogok stone tract of Burma, has also produced kornerupines, but in a rare and very attractive green colour.

Brown, and green, stones have come from Malagasy (Madagascar), while green and blue-green kornerupines were found early in the 1970s in both Kenya and Tanzania.

In 1975 new dark brown cat's-eye kornerupines came on the market. These were mostly small and opaque, with a well developed silvery streak and were perhaps the most attractive of the Sri Lankan kornerupines.

The absorption spectrum of kornerupine is by no means spec-

150

tacular and many specimens, indeed, show practically no distinct bands. But since the refractive index and density of the mineral are close to those of several other rare gemstones it is worthwhile knowing what to look for, if it is there. In good specimens a number of bands may be seen.

The best specimen we have seen so far as absorption bands are concerned is a large cut stone (9.12 cts.) in the British Museum (Natural History) collection, which we were allowed to examine. This stone was originally listed as idocrase and had been rescued from this false situation by Dr Herbert Smith, who assumed, much more plausibly, that the stone was an enstatite on the basis of its refractive indices. Being interested in the striking 506nm absorption band of enstatite, one of us obtained permission to examine the spectrum of this stone. This was done in the Keeper's room under poor lighting conditions, and, while it was obvious that the specimen did not show the narrow enstatite absorption band, not enough was seen or indeed known to suggest any other identity for the stone. It was not until some years later, when we were co-operating with the Mineral Department of the Museum in a study of kornerupines from Sri Lanka, that the true nature of this interesting specimen was realized.

The following measurements, then, were obtained from the museum specimen. These are divided into two series, one giving the bands seen in light which is vibrating parallel to the gamma direction, and the other being typical of the beta vibration direction, which is perceptibly different.

Gamma ray		Beta ray	
550nm	v.w.	540nm	v.w.
503nm	m.w.	503nm	m.w.
454nm	v.w.	463nm	v.w.
444nm	m.w.	446nm	s.
430nm	v.w.	430nm	v.w.

In the majority of Sri Lankan kornerupines the alpha ray is so dark a brownish red that when this is isolated by means of a polarizer it is difficult to see any bands at all beyond a rather broad one in the red. The whole field is too dim to enable this to be measured accurately, but it seems to be centred near 650nm. A similar band can be observed in the beta ray, which can be recognized by its green colour when seen through a polarizer. It is in this green ray that the 446 band shows quite clearly and strongly. In the gamma ray, which is yellowish-brown in tint, the 503 band is the only one clearly seen. Beyond this there is only a vague smudge in the blue, which can seldom be resolved into separate bands except in unusually favourable specimens such as the large museum stone. The spectrum is drawn in Fig. 23.1.

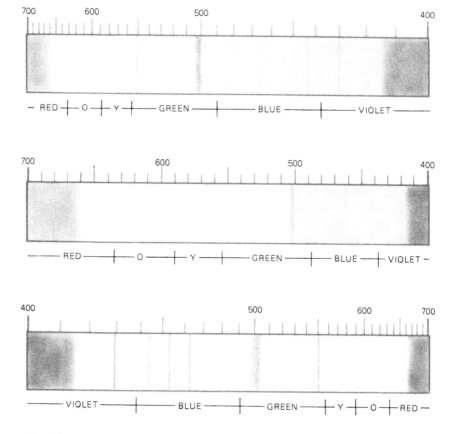

Fig. 23.1 KORNERUPINE. Not a strong spectrum but useful when it is seen

The attractive green kornerupines from Burma, and those from Malagasy and East Africa, show no definite absorption bands.

Axinite

Axinite is a borosilicate of aluminium and calcium, in which some of the calcium is replaced by ferrous iron, magnesium, or manganese. The magnificent bladed triclinic crystals of axinite found at Bourg d'Oisans and elsewhere are more suitable as mineral specimens than as a source of cut gemstones, but transparent material is sometimes sacrificed to please collectors of the rare and curious. The colour is a rather distinctive clove-brown, and cut stones are seldom free from flaws. Axinite has an absorption spectrum which is rather more attractive and definite than that of kornerupine. Three bands are usually pretty obvious at first glance, whatever the orientation of the specimen: a fairly compact and narrow band in the blue-green at 512nm and rather broader bands at 492 and 466nm in the blue. In favourable specimens another band can be seen at 444nm, and a strong and narrow band in the deep violet at 415nm. A very faint band at 532nm has also been detected. For reference, these bands can be listed in tabular form:

532nm	v.w.	narrow
512nm	m.	narrow
492nm	m.w.	broader
466nm	m.w.	broader
444nm	w.	broader
415nm	S.	narrow

Axinite is pleochroic, though not so strongly as kornerupine, and a polaroid disc may be useful in strengthening the bands should these be very weak. Viewing the stone on edge will also often yield stronger bands in this and in many other stones.

The drawing (Fig. 23.2) will help the reader to visualize the axinite spectrum.

Summary

Enstatite, kornerupine and axinite all have refractive indices and densities close enough to one another to make confusion possible unless care is taken to take critical and reliable measurements. The absorption spectra of the three minerals is thus of some importance to the gemmologist as an aid to discrimination. Enstatite has a completely distinctive spectrum, with its clear-cut line at 506nm in the blue-green, and other, less prominent bands. Kornerupine has no bands as definite as this, but a vague band at 503 may be noticed with a band or two in the blue which are difficult to resolve. In certain orientations, especially if a polarizer is used, a band at 446nm can appear quite strongly.

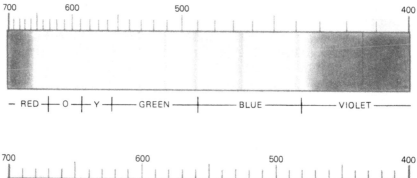

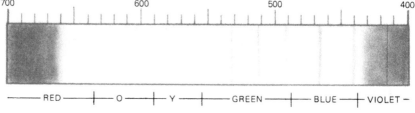

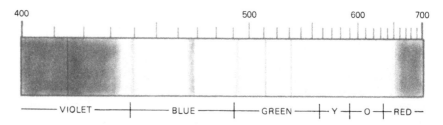

Fig. 23.2 AXINITE. Another rather vague spectrum which may help to identify this rarely cut gem

Axinite has essentially a three-band spectrum: one rather narrow band at 512 in the blue-green, and two broader bands not far away in the blue.

154

24   Absorption Spectra of Green Tourmaline and Iolite

Ferrous iron is also responsible for the deep greens and blues of tourmaline, and for the blue of iolite. These two minerals have little in common and are placed together here for convenience rather than for comparison. Some stones among them give recognizable absorption spectra.

Many of the allochromatic minerals (i.e., those which are colourless when 'pure') show a wide range of colour varieties, but, with the possible exception of corundum, none offers so great a range of colour as tourmaline. Red, pink, yellow, brown, green, blue, and black tourmalines are all quite often seen. Considering the numerous assemblage of elements packed away in the normal crystal of tourmaline, it is hardly surprising that the colourless variety (achroite) is comparatively rare. An attempt has been made to sub-divide tourmalines into different categories – *dravite*, or magnesium tourmalines, *schorlite*, or iron tourmalines, *elbaite* for those rich in alkalies, and so on. The pink and green tourmalines which are chiefly used in jewellery belong mainly to the alkali group, in which the composition can be roughly represented by the formula $H_8Na_2Li_3Al_{15}B_6Si_{12}O_{62}$. But, as we shall see, the elements which most influence the colour are iron and manganese.

A feature peculiar to tourmaline is the frequency with which one and the same crystal shows zones of different colour – usually pink and green. The different colour zones may be in concentric shells around the whole length of the crystals, or may be in segments parallel to the basal pedion. The divisions between the colours are commonly as sharp and distinct as in a strawberry-and-vanilla ice-cream block, and the causes for this curious particoloration have long been a matter for speculation and experiment. That the change in colour betokens a slight change in

155

composition is certain: in passing from a green zone to a pink one in a polished slab of this description a distinct fall in refractive index can be noticed, and the density of the pink tourmaline is also slightly lower than that of the green.

Dr J.E.S. Bradley and Dr Olive Bradley, of King's College, London, showed by X-ray diffraction[1] that the structure of the basic units of this complex mineral consists of rings of six tetrahedra of silica surrounded symmetrically by three BO_3 and three OH groups. The large centre spaces within the rings accommodate replacement elements sodium, calcium or fluorine, while iron and manganese, responsible for green and pink coloration respective, are attached at three sites around the rings which are themselves held together by atoms of aluminium.

The Bradleys compared the absorption of green tourmaline with that of ferrous ammonium sulphate, and that of pink tourmaline with manganous sulphate, and found close resemblances. Green tourmaline was found on analysis to contain more of both Fe and Mn than the pink variety, so that it is not easy to explain the mutual exclusiveness of the two colours or the dominant effect of manganese in pink tourmaline. But it was suggested that the answer must lie in the peculiar nature of the manganese atom, which only gives rise to colour under certain rather restricted conditions, but which, when these conditions are present, has a very intense colouring effect.

The few bands seen in green or blue tourmaline can be related to ferrous iron, but the more complex spectrum of pink or red tourmaline has little in common with the manganese spectrum of such minerals as spessartine or rhodonite.

Here we are concerned with the ferrous iron spectra of green, green-blue and blue tourmaline in which the main features are a rather sharp cut-off at the red end, down to 640nm or even further, a very faint absorption at about 560 and a ferrous iron band at 497nm, possibly with a much less obvious band at 462nm. (This band is quoted as 468nm in the original paper, and later as 462, which is the wavelength illustrated in T.H. Smith's original drawing. From this it would appear that 468 was a misprint which I have now corrected. Webster also quotes 468, but he would have taken this from Anderson's paper. Anderson quotes 462nm in his book *Gem Testing*). In dark brownish, green or blue stones the general absorption of the violet can obscure these bands and a polaroid filter to screen out the extraordinary ray may help to make them visible, as will a copper sulphate filter flask. In general the green tourmaline spectrum is regarded as not very diagnostic (Fig 24.1).

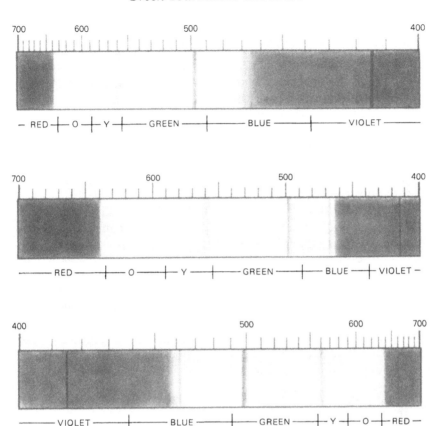

Fig. 24.1 GREEN TOURMALINE. Chiefly identifiable by the sharp cut-off at the red end, at 640nm or lower. The band at 415nm may be seen in deep blue tourmalines

Iolite absorption spectrum

The name *iolite*, which alludes to the violet-blue colour of this orthorhombic silicate, is a name more favoured by gemmologists than *cordierite*, which is preferred by mineralogists. Yet another name for the same mineral – *dichroite*, is a reminder of its most remarkable property – a pleochroism so striking that it is at once apparent when the stone is turned. The deepest ray, dark violet in colour, is seen in the direction of the optic axes, and belongs to light vibrating at right angles to the optic axial plane (corresponding to the beta refractive index). Since iolite is a negative mineral, the vibration direction of the fast ray (alpha index) is

157

parallel to the acute bisectrix: this ray is yellow or pale brown in colour. The slow ray (gamma index), vibrating parallel to the obtuse bisectrix, is of a clear blue tint. Only when viewed along an optic axis, where the deep violet ray alone is visible, or in the direction of the c axis of the crystal, when the blue and violet rays reinforce one another, are the colours seen by the unaided eye which make the mineral attractive as a gemstone. In other directions iolite has a most unattractive brownish yellow colour, with no hint of the blue or violet rays which must always be present even in such directions – as can be proved by the use of a polarizing plate which conjures up the more attractive tint when it is turned into the correct orientation.

Chemically, iolite is a magnesium aluminium silicate in which some of the magnesium is replaced by ferrous iron. The formula can be written $(Mg, Fe)_2Al_3Si_5AlO_{18}$. Some alkalies and water are often present.

In most faceted iolites the absorption will yield little that is exciting, but fine and good sized specimens will show a more rewarding spectrum which at first sight has a stormy look rather reminiscent of that of blue spinel. But there is little chance that they will be confused with each other when a tilt of the iolite to one side will at once show up its startling pleochroism.

The complete series of iolite bands is as follows:

645nm blue/violet
593
585
535   yellow
492
456
436
426   (near UV
398   (and
383   (UV
370   (regions

A polaroid filter will be needed to see the 645nm band and will probably enhance the rest of what is normally a rather ill-defined spectrum. The bands below 436 were recorded photographically and will not be seen, but are mentioned here for the sake of record. Bands in the ultra-violet have been noted in photographs of several other iron spectra, both ferrous and ferric.

The original article illustrated the spectra for the violet and

yellow rays separately, but they are so similar that I have drawn
the combined effect (Fig.24.2), which is normally seen.

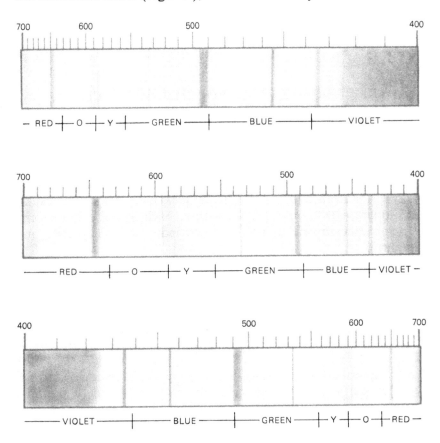

Fig. 24.2 IOLITE. A vague line spectrum seen in the yellow direction of
this strongly pleochroic gem. 645nm is very strong in blue direction

25 Absorption Spectra of Idocrase, Actinolite and Serpentine

Three ferrous iron spectra remain to be discussed although the absorption in each is scarcely distinctive. The minerals in question are idocrase, actinolite and serpentine, each of which have some resemblance to one or other of the jade minerals, jadeite and nephrite.

Actinolite, indeed, can itself be thought of as a true jade, since 'nephrite' is merely the term for a massive amphibole of the tremolite-actinolite series. 'Mutton-fat' nephrite is almost entirely iron-free tremolite, while the commoner green nephrite contains iron and is nearer to actinolite in composition. In recent years actinolite cat's-eyes from Taiwan have been sold as nephrite cat's-eyes.

Many minerals resemble jade in appearance and are used for similar purposes, and are often difficult to identify by simple means, so any help the spectroscope may give us is very welcome.

Idocrase

Idocrase or vesuvianite is a silicate of calcium, aluminium and magnesium containing a varying amount of iron and sometimes other elements. Though tetragonal, it is closely related in composition and structure to the garnets, and, like garnet, is usually formed by contact metamorphism, being found chiefly in altered limestones. The student of crystallography early becomes familiar with the attractive, well-formed brownish-green tetragonal crystals of this mineral. A clear, honey-brown variety from the Laurentian Mountains of Canada has provided attractive gemstones for the collector's cabinet. A massive grey-green translucent form known as 'californite' is rather similar in appearance to massive grossular ('Transvaal jade') and forms quite a

160

passable substitute for jade, though better known in the USA than in Great Britain.

The strongest and most constant band in the idocrase spectrum is centred in the blue at 462nm. Henri Becquerel, another of the famous family of French physicists, noted and measured this band as long ago as 1889, and further observed that it occurred only in the ordinary ray.

Brown idocrase, in common with all brown stones, absorbs extensively in the blue and violet regions, so that this band, strong though it is, can hardly be discerned in the general murk. In green types the band is broad and clearly seen, while in californite it appears as a narrower but still intense distinct band, well away from the general absorption. Careful survey of the spectrum of most idocrase specimens reveals the presence of other, far fainter bands. There is a narrow, weak band at 528nm with indications of a still fainter band on its long-wave side, and a vague and rather broad band centred near 487nm. Since all these bands belong to the ordinary ray, their strength can be much enhanced by viewing through a correctly oriented polaroid.

In some, but not all, of the brown Laurentian idocrases mentioned earlier there can be seen, in addition to these three bands, numerous fine lines in the yellow and green which we have learned to associate with the two rare-earth metals collectively known as didymium. These will be dealt with later, when rare-earth spectra are under discussion.

The idocrase spectrum is not very typical for ferrous iron – the most prominent band, for instance, is not situated near the junction of the green and the blue. It is thus possible that ferric iron here plays a prominent role. Trivalent iron is, in fact, known to be present, replacing some of the aluminium atoms.

The drawings (Fig. 25.1) show the strong 462nm band and the two fainter bands.

Actinolite

Actinolite has no absorption bands comparable in strength with the 462 band in idocrase: where this shouts, the feeble actinolite bands speak in whispers. There is a vague band in the yellow of undetermined wavelength, and two faint and narrow bands at 510 and 495nm. These bands, more particularly the 495 one, can sometimes be seen in specimens of nephrite. Mutton-fat nephrite seems to show in addition a vague band near 460nm. The presence of these bands, and the absence of the distinctive 437nm band of jadeite, may sometimes serve to identify nephrite and to distinguish it from its pyroxene rival.

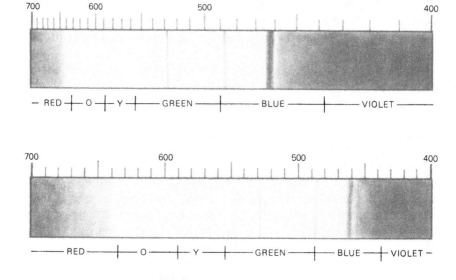

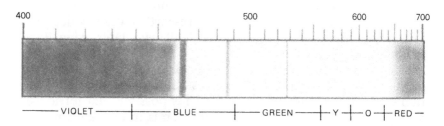

Fig. 25.1 IDOCRASE. The intense band at 462nm may give a cut-off of the remainder of the blue and violet

Serpentine

Serpentine is a name given to an extensive group of green monoclinic minerals which are essentially hydrous silicates of aluminium and magnesium or ferrous iron. The serpentines are closely related to another loose group of soft green minerals conveniently gathered together under the name 'chlorite'. Some of the serpentine minerals bear a close resemblance to jade, and thus come within the purview of the gemmologist.

Most of the translucent green serpentines show a typical

ferrous iron band at the junction of the green and blue, the measurements of which seem to vary a little between 492 and 498nm, and there is usually another band visible deeper into the blue, at 465nm or sometimes a little lower. There is a wide variation in strength both intrinsically and relatively of these two bands in the various sorts of serpentine.

The hardness of serpentine is very variable: usually it is between 3 and 4 on Mohs' scale and thereby easily distinguished from the jade minerals by a careful hardness test. The variety *bowenite*, however, is notably harder (up to 5½) and was in fact wrongly identified as nephrite by G.T. Bowen after whom it was named. Bowenite was first found in Smithfield, Rhode Island, USA, and from that locality is bright apple-green in colour, resembling nephrite. It is also known to occur at Milford Haven, in New Zealand, where the Maoris carve and use it in much the same way as the nephrite which is their most treasured material, and include it under the general term *pounamu*.

The bowenite most often seen in commercial use is reputed to come from China. It is usually a very pale greyish- or yellowish-green. Sometimes it resembles pale green jadeite in colour and translucency, but when one has seen a good deal of it the appearance is sufficiently distinctive to be recognized with some assurance at sight. A vague absorption band near 497nm has been observed in many specimens, and a rather stronger band near 464nm. It cannot be pretended that this spectrum has much diagnostic value, and the mineral's fairly constant density of 2.59 forms a far more reliable test where it is possible to carry out a determination.

Williamsite, which is the name given to a rich oil-green serpentine containing black octahedra of chromite, shows a more rewarding spectrum. There is a broad, vague, and weak band centred near 540, a narrow band at 495 is moderately strong and clear, and a vague band near 460nm is also visible.

Connemara marble is another serpentine material used extensively as an ornamental stone. A more famous name for the same type of material is *Verde Antique*, and the green jade-like pebbles from Iona are again much the same. These are marbles in which veins of rich green serpentine make a beautiful and varied pattern. This type of serpentine shows the 465nm band very strongly – though the strong general absorption may make it difficult to discern. There is also a band at 495nm, but this is weak and vague.

Another soft jade-like material may be mentioned here, though

163

it is more properly classified under the mineral penninite, of the chlorite group, than as a serpentine. This is often known as *pseudophite*, and in the trade had a certain vogue under the misnomer 'Styrian jade', presumably because it was mined in this Austrian province. Here, a narrow band at 498nm is seen at moderate strength.

Finally it may be mentioned that a specimen of the brownish-green serpentine known as *antigorite* was examined some years ago and gave a fairly distinct spectrum. It was in the form of a thin slab, and showed a pseudo-uniaxial figure. There was a vague band in the green, near 540nm, a narrow, clear band at 495, in moderate strength; a broad absorption region in the blue ending in a concentration at 452nm; and lastly a strong band at 430nm, forming a cut-off to the end of the spectrum.

Summary

The reader in search of spectroscopic data which will help in distinguishing between the many jade-like minerals may well feel that the data given in this chapter are of very little assistance to him. One must admit that, in jade identification, only the powerful line at 437nm in jadeite is sufficiently constant and distinctive to be of first-class diagnostic value. If, on grounds of appearance, or as the result of hardness, specific gravity or refractive index tests, idocrase of the californite variety is suspected, the intense band at 462nm in the blue, described and illustrated above, will form a useful confirmatory test, and so will the rather similar band if it is seen in a mottled green and white marble, help to confirm the stone as verde antique serpentine. But the other bands mentioned in connection with actinolite and serpentine are too vague and variable, and too lacking in distinction one from the other, to have any real practical significance, and they are therefore not illustrated.

Scientifically speaking, it is interesting to note how the region near 500nm is a key position for spectra due to ferrous iron.

26 Absorption Spectra of Sapphire and Chrysoberyl

We now come to absorption spectra due to ferric iron – that is, trivalent iron corresponding to the oxide Fe_2O_3. Whereas ferrous (divalent) iron is mostly found in magnesium minerals due to partial isomorphous replacement of MgO by FeO, in the minerals we shall now discuss the molecule Fe_2O_3 is commonly found replacing alumina (Al_2O_3). Less than 1 per cent of ferric oxide may suffice to influence the colour greatly and produce strong absorption bands.

With ferric iron, the strongest bands are of shorter wavelength than with ferrous iron, being found in the 450nm region, where the blue merges into the violet.

Absorption bands in sapphire

The spectrum we shall be describing belongs properly to *green* sapphire, which owes its colour entirely to ferric iron. But the same spectrum is present to some extent in natural blue sapphires also. *Synthetic sapphires, whether blue or green, show none of these bands*: hence the supreme importance of these effects to the practical gemmologist.

In green sapphires from any locality three evenly spaced absorption bands can be seen in the blue region, centred at 471, 460 and 450nm. The 450 band is by far the strongest of these, and almost merges with the next strongest, the 460 band. The 471nm band can be seen as distinctly separate from the block formed by the other two, and it is the least strong of the three. Of blue sapphires, Australian stones, which are rich in iron, have the three absorption bands very strongly developed. At the other end of the scale, Sri Lankan sapphires show only the 450nm band, and that very faintly, while sapphires from Burma, Kashmir, Thailand and Montana show the 450 band very clearly and usually a 'smudge' to the longwave side of this to indicate traces of the other two absorption bands. In cases where the 450 band is excessively weak, the use of a flask of copper sulphate solution as a condenser in place of the usual plain water flask will make the line more easily discernible, as the glare due to the other spectrum colours is muted. The band can also be strengthened by turning a polaroid disc over the eyepiece of the spectroscope until the position of maximum strength is obtained. All three bands belong to the ordinary ray.

In deep-blue natural sapphires there is often an ill-defined broad absorption centred at about 585nm. This, and the main absorption bands in the blue are shown in the drawings (Fig. 26.2). The green sapphire spectrum, with the three-band group at full strength, is shown in separate drawings (Fig. 26.1).

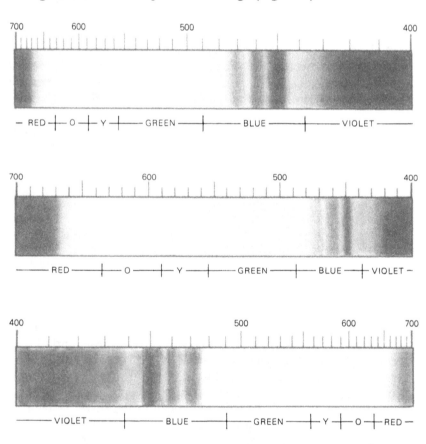

Fig. 26.1 GREEN SAPPHIRE. Three intense absorption bands in the blue are due to ferric iron. The second and third of these may fuse to form a solid block

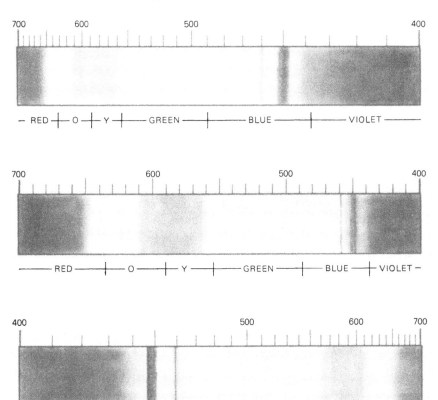

Fig. 26.2 BLUE SAPPHIRE. A residue of the green sapphire absorption is to be seen in most blue sapphires, although those from Sri Lanka may show it only very faintly. It may also be seen in some yellow sapphires

Yellow sapphires also show the bands. In those from Australia, Siam, or Montana they are there in fair strength. In Sri Lankan yellow sapphires, however, only the 450 band can be discerned, and in small specimens even this may not be visible. It may be remembered that, as a compensating factor from the diagnostic point of view, Sri Lankan yellows have a characteristic apricot-coloured fluorescence both in long- and short-wave ultra-violet light. Thus we have a useful means of distinction between natural and synthetic yellow corundum in cases where the microscope yields no definite signs. Stones which show one or more of the absorption bands in the blue or show an orange fluorescence are almost certainly natural sapphires: those which show neither

167

effect are synthetic. One says 'almost certainly' as a precaution, since Webster found an occasional synthetic yellow sapphire which shows a fluorescence somewhat resembling that of Sri Lankan stones.

Bands in ultra-violet

Absorption of sapphire in the ultra-violet also shows distinctions between stones from different localities and between natural and synthetic. Photographs of the spectrum have shown the presence of strong absorption bands in the near ultra-violet at about 379 and 364nm respectively. These are almost certainly due to ferric iron, and can be seen quite strongly even in Sri Lankan sapphires where the 450 band is very weak. In stones in which the bands in the blue are well developed, the ultra-violet bands are swallowed in a general absorption region covering all the near ultra-violet. The important point is that synthetic sapphires show neither the visible nor the ultra-violet absorption bands. At first sight this is a puzzling fact, since iron oxide as well as titania is added to the powdered alumina as raw material for the production of synthetic sapphires by the Verneuil process. But, though the addition of iron seems to influence the final colour of the boule, the oxide itself (being of lower melting-point than titania) appears to evaporate in the extreme heat of the furnace and, in Prospero's phrase, 'leave not a rack behind'.

Chrysoberyl

We have already described the spectrum of alexandrite chrysoberyl under the heading of chromium spectra – but there are commoner, and in some ways more attractive, golden yellow and greenish chrysoberyls which contain no chromium and owe their colour entirely to iron, which replaces a little of the alumina in this beryllium aluminate. The result is a strong, broad band centred at 444nm, which can be distinguished from the sapphire bands not merely from its position but in being a single solid block instead of a threefold complex. In some of the deeper-coloured chrysoberyls, two faint, narrow bands are seen at 505 and 485nm, and these are shown in the drawings (Fig. 26.3). In brown chrysoberyl, which resembles tourmaline in appearance, the general absorption of the blue and violet masks even the 444 band.

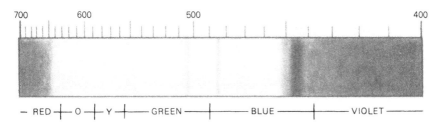

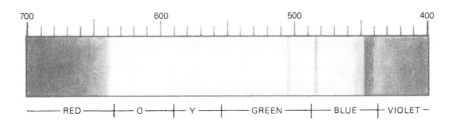

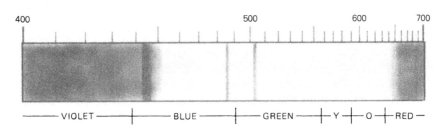

Fig. 26.3 CHRYSOBERYL. The powerful absorption band at 444nm identifies all but the alexandrite variety

The honey-yellow to greenish-brown chrysoberyl cat's-eyes, which are amongst the most beautiful and valuable of gem materials, show the 444 band strongly where enough light can be punched through them, and this forms a useful proof that they are indeed chrysoberyl and not one of the far less valuable quartz cat's-eyes which can sometimes so closely resemble them. In the case of an unmounted cat's-eye a simple test in heavy liquid of course forms an even more convincing means of separating the two species, but the spectroscope is applicable even for mounted stones where a density test cannot be carried out.

27 Absorption Spectra of Aquamarine and Orthoclase: Spodumene and Jadeite

Aquamarine and yellow orthoclase also owe their very different colours to ferric iron and have absorption spectra which are somewhat similar. They have been placed together for that reason, although the blue or blue-green of aquamarine is unlikely ever to be confused with the yellow of the orthoclase.

In small specimens of aquamarine it is difficult to see any bands of sufficient strength to measure. But aquamarines as used in jewellery today are often rather large, and in these a somewhat broad band in the violet at 427nm can be seen in moderate strength. Another band can be seen in the blue-violet at 456nm. This is feeble and diffuse, but serves to form a 'pattern' with the stronger band which the experienced observer can recognize and which is a fairly conclusive test for this popular gemstone. Actually, the most frequent imitations of aquamarine are either pale blue synthetic spinels, which show a distinctive three-band absorption due to cobalt, or simply a low refractive-index glass of appropriate colour, which shows no bands of any kind.

In aquamarine the stronger colour belongs to the extraordinary ray; in fact, the ordinary ray is practically colourless. The absorption bands are also stronger in the 'extraordinary' spectrum, and they can thus be enhanced by turning a polaroid disc either above the eyepiece or below the slit of the spectroscope until the optimum position is reached. When the spectrum of the extraordinary ray is thus isolated an exceptionally narrow and delicate absorption line may be detected in the middle green at 537nm. This line is elusive and difficult to see even when it is present, so focus and slit width need to be exactly right. It has also been seen in some yellow beryls and even in some colourless ones. Where it occurs it can be a useful check, particularly for rough specimens which cannot be tested by refractometer. The drawings (Fig. 27.1) show the two broader bands and the 537 line, the latter with rather more emphasis than this exceedingly thin line has in reality.

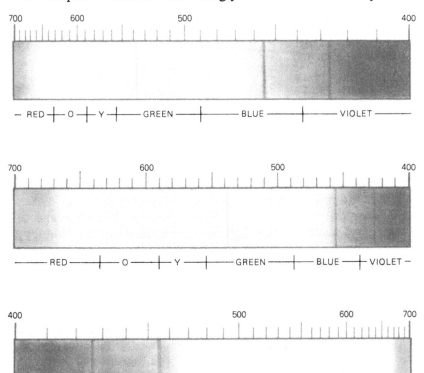

Fig. 27.1 AQUAMARINE. Narrow bands in the blue and the extremely narrow faint line at 537nm are seen only in the extraordinary ray. Use polaroid

It would be rash to ascribe the 537 band to ferric iron without careful experimental proof, as its character is so unlike the normal broad absorption bands of iron. On the other hand it is difficult to think of any other element which is commonly present in beryl in small traces and which could be expected to produce such a line.

Maxixe beryl

A deep blue beryl found early this century at the Maxixe Mine in Minas Gerais, Brazil, has a most unusual spectrum, totally different from anything seen in other naturally coloured aquamarines, and due either to rare earth elements or to uranous uranium contaminants. Much later a very similar material was identified as having been artificially irradiated. These will be discussed in detail later, under 'Rare earth spectra.'

Yellow orthoclase

Orthoclase, the monoclinic potash feldspar, is a major constituent in the commonest igneous rock, granite, and is thus a very important rock-forming mineral. In its pure form it is completely colourless, and its low refractive index and lack of 'fire' leave it unattractive as a faceted gemstone. In the moonstone variety this is offset by the beautiful bluish-white schiller that this gem displays. Neither moonstone nor colourless transparent orthoclase show any absorption bands, but the yellow orthoclase from Madagascar (Malagasy) contains some 5 per cent of ferric iron replacing some of the alumina in the orthoclase molecule, and this variety shows absorption bands in the blue and violet very similar to those just described in aquamarine. There is a weak band at 448nm accompanied by a stronger band centred at 420nm: both are rather broad and diffuse. Photographs of the yellow orthoclase spectrum have recorded a further and much stronger band in the near ultraviolet at approximately 375nm. This band is about 15nm broad. No similar band has been detected in photographs of the aquamarine spectrum, while on the other hand no trace of a narrow line in the green as seen in beryl has been noted in the orthoclase spectrum.

Jadeite and yellow spodumene

The absorption of jadeite has been described and illustrated when we were dealing with chromium spectra, and the similarity of the iron bands seen in this gem and in yellow spodumene was briefly mentioned. The ferric iron absorption bands are in fact practically identical in these two gem minerals. Both minerals are monoclinic pyroxenes; jadeite is a sodium aluminium silicate, while spodumene is a lithium aluminium silicate of analogous composition. So it is hardly surprising that the absorption effects are similar. If the spectroscope were used without having seen the colour of the stone tested it would indeed be impossible to separate them, for the two bands in the far blue are at almost the same wavelengths in both gems. The more popular lilac coloured spodumene known as kunzite does not have iron impurities and so does not show the bands. Spodumenes of a pronounced yellow or yellow-green are the most likely to show this absorption. The stronger of the two bands is at 437nm and its weaker companion forms a sort of echo at 433nm.

P.L. Bayley was the first to record the two violet bands in spodumene[1]: we have seen no references in the literature to the corresponding bands in jadeite prior to our own observations. A pale yellowish crystal of spodumene from Burma showed a faint narrow band at 505nm, similar in appearance, though weaker, to bands in this position in enstatite, diopside and diamond.

In jadeite, particularly of the paler types, the 437 band can nearly always be seen in a small spectroscope by reflected light, and this forms a most valuable check on the authenticity of 'jade' ornaments and earrings. In dark green specimens the band is probably always present but may be very difficult to detect in the general absorption due to chromium. This is a case where the slit of the spectroscope may profitably be widened considerably to enable one to glimpse the band. Use of one of the blue filters may also be helpful.

Fig. 16.1 shows the narrow bands in the violet.

28 Absorption Spectra of Andradite and Epidote

The number of gem minerals which show absorption bands due to ferric iron are far fewer than those in the ferrous iron series: andradite and epidote are indeed the only remaining species in this category that we need to describe.

Andradite

Andradite is the mineralogical name for the calcium-iron garnet of which the beautiful green demantoid is the main gem variety. The absorption of demantoid has already been described when dealing with the chromium spectrum series, since the rich green colour which contributes to the attractiveness of this gem is mainly due to chromium, and typical chromium lines in the red and orange are always to be seen in fine specimens. But not all gem andradites are green, and the chromium lines are thus not so reliable a guide to the garnet as the bands in the blue and violet which, being due to ferric iron (an essential constituent) are a constant feature of this mineral.

The strongest of these bands is a very intense and fairly narrow one in the blue-violet, centred at 443nm. In deep green specimens there is a general absorption of the violet due to chromium, so that this band can only then be seen as a sharp and complete cut-off to this end of the spectrum. In the paler and yellower types (sometimes, but not very advisedly called by the name 'topazo-lite') the band stands out clearly with the violet visible beyond, and forms a very satisfying sight, being intense, sharply defined, and fairly narrow.

Two other and far fainter bands are sometimes to be seen at 485nm and 464nm, but these have small diagnostic importance. The drawing (Fig. 17.1) shows the spectrum of a green demantoid in which chromium bands in the red are visible in addition to the powerful 443nm iron band, which is nearly swallowed up in the general absorption of the violet.

Epidote

The absorption spectrum of epidote was one of the first to be examined in any detail, having been studied by H. Becquerel in 1889[1], when he discovered that each of the three rays corresponding with the alpha, beta, and gamma refractive indices had a different spectrum, and gave measurements for the bands seen in each case.

In its crystal form epidote is often described as 'pistachio green' in colour, a characteristic of the mineral which leads to it being known also as 'pistacite'. The pistachio nut is used in confectionery, but few will be familiar with its velvet-green outer husk, which resembles the granular outer coating of many epidote crystals and accounts for this somewhat outlandish name. However, many crystals are found without this green coating and fine lustrous dark brown or black specimens are to be seen in most mineral collections. The cut stones are generally a dark greenish-brown.

As with most brown or brownish stones, there is heavy absorption of the violet, and consequently the very intense band near 455nm which is the most constant feature of the epidote spectrum is only clearly seen in exceptionally pale and transparent specimens. Epidote, to be truthful, is vastly more attractive in its natural lustrous monoclinic crystal form than when cut as a gemstone; but cut stones are sought by collectors.

The position of the strong band varies somewhat with the vibration direction of the light, as Becquerel found. He gave 4580Å (458nm) as the measurement in the alpha ray, and 4570Å (457nm) for the beta ray. In the gamma ray it is virtually absent. In the beta ray another much weaker band can be seen at 475nm.

A clear green specimen of the closely related mineral clinozoisite from Kenya, which we examined some years ago, showed a very similar spectrum to that of epidote, with faint chromium lines in the red in addition.

In retrospect

In brief retrospect, it may be said that the absorption spectra due to ferric iron tend to show a strong absorption band in the blue-violet, accompanied by weaker bands on the long-wave side of this. In only four cases (sapphire, chrysoberyl, jadeite and demantoid) can the bands be said to have first-class diagnostic importance. Undoubtedly the most important spectrum of all is that of sapphire, since it enables the observer to distinguish between natural sapphires and synthetics, which show no trace of the 450nm sapphire band.

29 Absorption Spectra Due to Manganese

There are three gem materials which contain manganese as part of their essential composition and thus rank as idiochromatic minerals. These are the manganese silicate, rhodonite; the manganese carbonate, rhodochrosite; and the rare manganese garnet, spessartite or spessartine.

The first two are best known in massive form as translucent ornamental minerals having an almost identical rose-pink colour – but transparent specimens of each are sometimes encountered, and rhodonite in this case may show a colour which approaches more closely to the 'aurora-red' or orange-brown tint which is most typical of spessartite garnet.

In each of these minerals manganese is present in the manganous or divalent state. Absorption curves of manganous salts are known to show peaks of maximum absorption in the green and blue-violet parts of the spectrum. This is true also for both rhodonite and rhodochrosite, in each of which a broad absorption band can be seen in the green, a rather vague band in the blue and (in suitable specimens) an intense band in the violet. Rhodonite and rhodochrosite are so similar in colour and general appearance, density and mean refractive index, that it would be pleasant to be able to state that the spectroscope provides an easy means of distinguishing between them. But though differences in detail may be detected in exceptional specimens it must in honesty be admitted that the spectra are so similar in the visible region that it would be unsafe to base any decision on absorption bands only. This is in fact one of the rare cases in gemmology where a hardness test is perhaps one of the simplest and surest aids to identification. Rhodochrosite has a hardness of only four on Mohs' scale and can easily be scratched by a knife, while rhodonite has a hardness near six, and a knife in consequence makes little impression. In passing it may be remarked that

177

another test which one might expect to be infallible where a carbonate is to be distinguished from a silicate, i.e. the application of a drop of hydrochloric acid to the surface of the stone, and watching for the effervescence of carbon dioxide evolved, is here by no means foolproof, since rhodonite may contain enough carbonate to show perceptible effervescence also.

In the common translucent massive forms of the two minerals, often only the broad band in the green can be clearly seen by transmitted or reflected light. In rhodonite this is centred at about 548nm and in rhodochrosite near 551nm: in bands so vague this is not sufficient difference to separate the two with any certainty. If a strong enough beam of light can be transmitted, the broad bands in the blue can also be discerned, though often they may merely

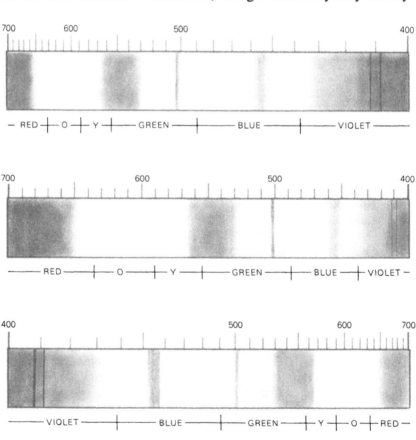

Fig. 29.1 RHODONITE. A manganese spectrum seen best in the rare transparent version of this mineral

178

form the beginning of the general absorption extending to the end of the spectrum. For record purposes, a full list of bands as measured by us in transparent specimens of each mineral can now be given.

Rhodonite

A transparent specimen from Australia, having density 3.71 and refractive indices 1.733 and 1.746, showed the following bands: 548nm, broad and moderately strong; 503nm narrow and moderately strong; 455nm, broad, vague, and rather weak; 412nm and 408nm, strong and rather narrow, which could just be detected in the deep violet (Fig. 29.1).

Rhodochrosite

Transparent rhodochrosite from Colorado, having density 3.69 and refractive indices 1.60 and 1.82, showed the following absorption bands: 551nm, moderate, broad; 454nm, broad and rather vague; 410nm (approx.), narrow and intense. A photograph of this spectrum showed further strong bands in the near ultra-violet at 391, 383, 378 and 363nm (Fig. 29.2).

Spessartite

The spectrum of spessartite is of considerable practical value in identifying this rare and often beautiful garnet. When pure, its refractive index and density are very near those for almandine, and, though the difference in colour should usually distinguish between them, there is often enough of the almandine molecule present to impart a port-wine tint to the garnet, or give it the hue of a hessonite, and the distinctive manganese bands in the spectrum form a welcome confirmation in such cases. When the main almandine bands are also present, as they often are, the spectrum may be somewhat confusing for the beginner. The specifically spessartite bands begin with a weak one at 495nm, followed by a vaguer band centred near 489nm, a stronger one at 462nm, and a powerful band at 432nm. This is often the last one visible, but narrow bands at 424 and 412nm (the latter very intense), have also been measured, while finally two at 406 and 394nm were detected in a photograph of the spectrum. It will be noted how the wavelength 412nm is a key position for powerful bands in all the three manganese spectra described (Fig. 29.3).

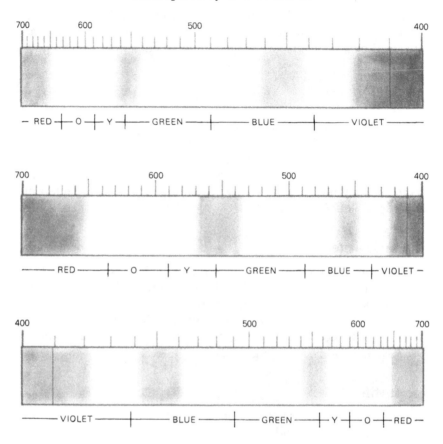

Fig. 29.2 RHODOCHROSITE. A vague spectrum when compared with rhodonite. The intense band at 410nm is not easily seen at the extreme edge of human vision

Red tourmaline

It has been mentioned that research work by the Bradleys gave strong support to the suggestion that pink and red tourmalines owe their colour chiefly to the presence of manganese. While the absorption curves seem to show that this is true, there is little in common between the absorption bands seen in the interesting spectrum of pink or red tourmalines and those just described for the manganese minerals. The spectrum bands are actually most strongly displayed in tourmalines of rather brownish red tint. There is a broad absorption region in the green, usually centred

180

near 525nm, within the long-wave end of which is seen a very distinctive narrow line at 537nm. There are in addition two bands in the blue which are almost as narrow as the blue lines in ruby. These are at wavelengths 458 and 450nm. These narrow lines in the blue can be seen in many pink or red tourmalines which do not show the peculiar broad band enclosing a narrow line in the green (Fig. 29.4).

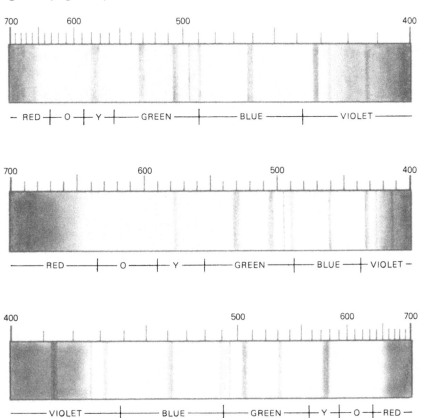

Fig. 29.3 SPESSARTITE. A combination of faint ferrous iron spectrum of almandine and the manganese spectrum of spessartite. Pure spessartite is rare

Willemite

Among the rarer gem minerals beloved of collectors, willemite gives a further example of a manganese absorption spectrum which is to an extent analogous to that of spessartite garnet, although the gems themselves are quite dissimilar. Here weak bands at 583 and 540nm are followed by a vague one at 490 and two clearer ones at 442 and 432nm in the violet, ending with a very strong absorption at 421nm. Willemite fluoresces a bright green under ultra-violet light and could be confused with a similarly coloured yellow-green synthetic spinel, which has very similar fluorescence.

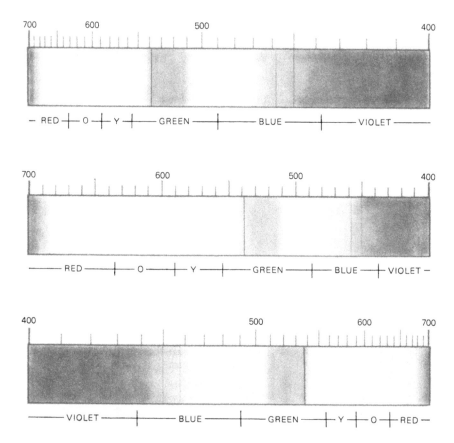

Fig. 29.4 RED TOURMALINE. Another manganese spectrum. The sharp line in the broadly absorbed green is the main identification

Synthetic green spinel

The spinels have two narrow bands at 448.5 and 423nm, which were first spotted by Robert Webster. The 423 one is the narrower and stronger of the two.

182

Manganese

The same bands may sometimes be seen in addition to cobalt bands in darker blue spinels imitating zircon or aquamarine, where manganese has probably been added to induce a greenish cast to the stones to improve their resemblance to the real gems they are imitating.

30   Absorption Spectra Due to Cobalt and Vanadium

The use of cobalt as a blue pigmenting material for glazes and decoration of pottery was practised by the Chinese as early as the T'ang dynasty (AD 618-906), and earlier than this in Persia. From the sixteenth century onwards cobalt ores have been regularly used in the production of blue glass, though it was not until 1735 that the metal was isolated and named by the Swedish chemist, Brandt.

We are, in fact, accustomed to thinking of cobalt in terms of *blue*. It seems strange, therefore, that very few of the naturally occurring cobalt minerals and none of the crystalline salts of cobalt are blue in colour, but are found either as ores of metallic lustre or as pink or red compounds. Hantzsch, who wrote a classic paper on the colour of cobalt compounds (published in 1927) was the first to find the clue to this curious behaviour, and subsequent workers have confirmed his findings. It has been established that pink or red colours are found where the cobalt ion is surrounded by six oxygens, while blue colours are produced when the ions are surrounded by only four oxygens.

For example, if periclase, the cubic form of magnesium oxide, is heated with cobalt nitrate, some of the magnesium atoms are replaced by those of cobalt and a red colour results, since in the periclase structure each magnesium atom is surrounded by six oxygens; whereas if spinel is the host lattice, where each magnesium is in contact with only four oxygens, cobalt gives rise to an intense blue coloration. Although almost all natural blue spinel owes its colour to ferrous iron and not to cobalt.[1]

Synthetic blue spinel, on the other hand, almost always owes its colour to cobalt, and so do many of the blue glasses used as imitation gems. Blue cobalt glass is also used very often as the base for garnet-topped 'sapphire' doublets. Thus we have three types of counterfeit gems (not to mention certain plastics) which display the very typical cobalt-blue spectrum, and its presence gives useful warning that we are dealing with some sort of fake.

The absorption curve of red cobalt compounds rises from a minimum in the deep red to a single peak in the green at about 520nm – falling to a low value again in the violet. Cobalt blue curves vary in detail, but agree in showing three distinct maxima in the orange, yellow and green. Absorption between these maximum peaks remains high, so that there is strong absorption for the entire region between about 670 and 520nm, with notable transparency not only in the blue and violet (as one would expect) but also in the deep red (Fig. 30.1).

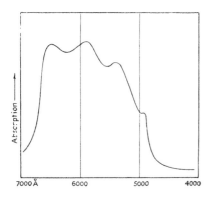

Fig. 30.1 Absorption curve of
COBALT GLASS

The strong absorption of the yellow and transmission in the red makes cobalt glass a useful filter for detecting the red flame of a potassium salt volatilized in the flame of a bunsen burner even when the presence of sodium would drown the effect with a blaze of yellow to the unscreened eye. Cobalt glass goggles are also preferred to other protective goggles by men who have to view furnaces, since a rapid change in colour sensation is noticeable for only a slight change in the furnace temperature. One further example of how the absorption bands in cobalt glass have been utilized was seen in the last war, where it was found that cobalt glass windows or even skylights could form an effective 'blackout' when sodium light was used to illuminate the workshop or factory within.

The strong transmission in the deep red and absorption of the green which is so characteristic of blue cobalt colours causes them to appear strikingly red through the Chelsea filter, which can pass only deep red and yellow-green. The curious red flashes which can be seen in some synthetic blue spinels under artificial light are

185

another instance of the effects caused by this transmission of deep red by cobalt-blue media.

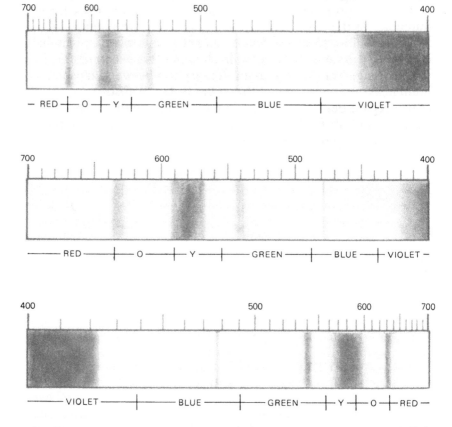

Fig. 30.2 SYNTHETIC BLUE SPINEL. The three band spectrum of cobalt is another easily recognized pattern. Here the central band in the yellow is broadest and most strongly absorbed

Synthetic blue spinel

As with most absorption spectra the effect seen through the spectroscope will increase with the depth of colour of the stone tested. In deep blue cobalt synthetic spinels the absorption may extend in a solid block from about 670 to 510nm. But stones of moderate 'sapphire' blue will resolve the block into three strong bands centred in the orange at 630, in the yellow at 580 and in the green at 543nm. Of these the middle one is the broadest and darkest while that at 543nm tends to be the faintest, narrowest and

186

vaguest. A weaker and narrower band at 478nm may also be seen in some darker stones but this is not significant for identification.

Spinels made to resemble aquamarines will show the main bands only very faintly and 543 may be difficult to see. Fig. 30.2 shows the bands as seen in a stone of medium blue colour.

It may be noted that some blue synthetic spinels have been made which do not have cobalt as the colouring element. These, of course, will not give a cobalt spectrum and are usually inert to this test.

Cobalt glass

The position and relative intensities of cobalt absorption bands in glass vary with the nature of the host material. However the spectrum is unmistakably a 'cobalt' one and differs far more from that of cobalt spinel than from those of other glasses. Typical measurements for the centres of the three main bands are 656 in the red, 590 in the yellow and 538nm in the green. They are therefore more widely spread in glass than in spinel and in addition the centre one is now markedly narrower than the other two while the 656nm is noticeably the darkest. See Fig. 30.3.

Another very vague and weak band may occasionally be seen at 495nm but this has no diagnostic significance.

In cobalt-blue doublets the absorption pattern will be as for glass, with perhaps the strongest almandine band at 505nm showing faintly as a contribution from the garnet top.

Cobalt dyed chalcedonies

Some cloudy white chalcedonies have been stained to an intense blue, or a less intense purplish blue with cobalt salts. These show a strong cobalt absorption spectrum in which the longest wavelength band is the broadest and markedly the most intense. 645, 580 and 520nm are the approximate centres, so it seems that they are spread rather as in glass, but have moved further towards the blue in the cryptocrystalline conditions of this host. Paler stones may show only vestigial traces of these bands. There is some suggestion that this stain may fade with age and long exposure to light.

Among the chalcedonies tested an exceptionally deep and saturated blue was found to be a fine purple-red by transmitted light and gave an intense transmission in the red and a much fainter one in the blue, all else being totally absorbed. Another dark stone did not give the red transmission.

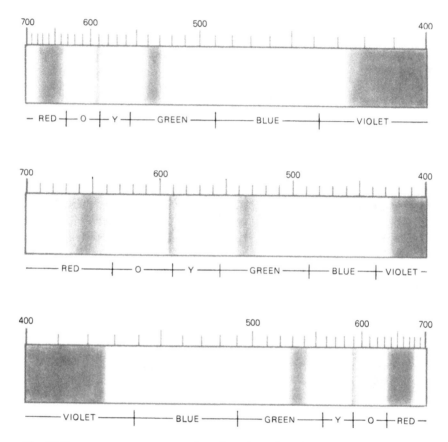

Fig. 30.3 COBALT BLUE GLASS. Another three band spectrum very similar to 30.2, but here the bands are more widely spread, although varying with the type of glass, and the left-hand one in the red is now the broadest and darkest

Plastics

Synthetic resins or plastics are sometimes coloured blue with cobalt, and yield a typical cobalt spectrum. Such fancy materials are hardly likely to be confused with glass or synthetic spinel, but it is perhaps interesting to note that once again the nature of the host material affects the position of the absorption bands. A cobalt-blue sample of polystyrene, for instance, gave the measurements 652, 610, and 566nm for the centres of the three main bands.

Spectra due to vanadium

Cobalt is scarcely ever found as a colouring agent in any natural gemstone. The same can be said of vanadium. A colour-change natural sapphire from Burma was found to show a vanadium spectrum, but by far the commonest instance of a vanadium colour and spectrum in gemmology is provided by that strangely popular synthetic corundum which is made to simulate alexandrite. This has a curious plum colour or slaty bluish tint in daylight, and a rich purple red in artificial light. To the accustomed eye it looks like nothing except itself – perhaps the easiest of all the synthetics to spot at sight. It is also usually very obliging in revealing strong curved growth striae. However, the single absorption line due to the vanadium used in its manufacture is a useful additional check. This is in the blue at 475nm, and is exceptionally clear and sharp, though usually not very strong. The rare form of natural sapphire which also shows this line has quite a different appearance and can hardly be said to invalidate this test.

Some of the synthetics have a little chromium added with the vanadium, presumably to enhance the red effect. This is revealed by a rather weak fluorescent line at 693nm.

31   Absorption Due to Copper: The Spectrum of Turquoise

From a casual knowledge of familiar copper salts (e.g., copper sulphate) and of the idiochromatic copper minerals used for gem or ornamental purposes (turquoise, malachite and azurite), we gain the impression that copper always gives rise to shades of blue and green. This is further borne out by blue glasses and ceramic glazes, also based on copper, which have been known and admired from the days of ancient Egypt and Persia.

But it must be remembered that these are all products of divalent (cupric) copper. The compounds of monovalent (cuprous) copper are more often deep red in hue, and the famous 'Sang-de-Boeuf' glaze was produced by suitable reduction of cupric compounds in the melt to a lower state of oxidation. Cuprous oxide (Cu_2O) is found in blood-red octahedra as the mineral cuprite, though the colour is so deep and the surface of the crystals so often coated or tarnished that the true magnificence of the colour can seldom be appreciated in the hand specimen. When a suitable crystal is detached and a beam of light transmitted through it, the spectroscope reveals quite an astonishing absorption spectrum. The entire spectrum is swallowed up by an intense and continuous absorption region from the violet to the orange, after which the absorption ends abruptly in an edge of razor sharpness at about 630nm. Thus only a narrow band of red light reaches the eye. Cuprite is indeed a 'spectrum filter' transcending in efficiency even the excellent red gelatine filters supplied by photograhic companies.

The deep red 'copper ruby' glasses probably contain minute crystals of elementary copper, and do not owe their colour to cuprous ions.

Copper blues and greens

While cuprous copper may transmit only red light, as we have seen, cupric compounds are the most efficient *absorbers* of red light that we know. Hence copper blue or green glasses are in great demand for signal purposes, since their strong absorption of the longer waves precludes the risk of their being mistaken for colourless or red in foggy weather. The influence of the co-ordination (i.e. the number of atoms surrounding the cupric ion) is clearly important. Thus crystals of 'blue vitriol', i.e. cupric sulphate, when heated strongly enough to drive off their five molecules of water of crystallization, lose their brilliant blue colour and are reduced to a colourless powder, while malachite and azurite, which are both basic carbonates of copper, owe their different colours to a variation in the content of hydroxyl ions.

Malachite and azurite are both too opaque for their spectrum to be studied by transmitted light, and by reflected light no definite bands can be seen.

The spectrum of turquoise

There remains the hydrous phosphate of aluminium and copper known as turquoise, which has been valued as a gemstone since the beginnings of civilization. And turquoise, most fortunately for gemmologists, does reveal absorption bands. These are not at all obvious, and were, in fact, not discovered until 1940, when they were observed, measured and photographed in our laboratory.

An absorption band of moderate width and strength was first noticed in light transmitted through the thin edge of a cabochon of Persian turquoise. It was expected that Egyptian turquoise, which is notably more translucent than other types, and very well coloured, would show the band, which was in the violet at 432nm, more readily – but this was not so. But photography subsequently revealed in light passed through specimens from both localities, not only the original 432 band, but another band of similar breadth and strength near 420nm. This band is very seldom visible to the eye owing to the general absorption in this region of the violet. But on the long-wave side of the 432 band a broader, fainter absorption band is nearly always visible, centred at 460nm. The two bands 460 and 432, taken together form quite a distinctive spectrum, and are indeed specific for turquoise, since none of the other minerals at all resembling turquoise, and none of the many imitations of the stone, has been found to show bands anything like these.

191

If one had to rely on transmitted light for viewing the turquoise spectrum it would have small practical value, for few indeed of the turquoise pieces used in jewellery can be made to pass more than a glimpse of light even from a concentrated 500 watt source. But fortunately, as with jadeite, the spectrum can be seen very well by light reflected from the surface of the specimen, particularly if a blue filter can be used to cut out glare from the longer wavelengths.

Specimens of turquoise from all known (and many unknown) localities have shown the weak 460 and stronger 432 bands, and the spectrum is undoubtedly very useful indeed as a means of testing a mineral on which so few tests can satisfactorily be performed

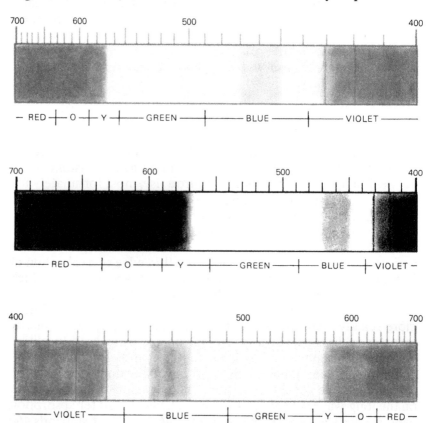

Fig. 31.1 TURQUOISE. A spectrum obtained by reflecting light indirectly from an opaque surface. The line at 433nm should be visible but 420 may be lost in the low sensitivity of the eye to the violet wavelengths

and which can be so closely simulated by a wide range of natural and artificial substances. Undoubtedly a strong source of light is essential – and one may here utter a warning not to concentrate light on to any turquoise through a bullseye or other glass condenser, unless the heat rays have first been filtered out by passing through water, since overheating this mineral may easily cause cracking or discoloration. It was an accident of this kind that originally led us to adopt as our standard practice the condensing of our 500-watt light through a water flask – a method which is simple, efficient, and safe.

The drawings (Fig. 31.1) will help the reader to visualize the appearance of the turquoise spectrum. As mentioned above, the 420nm band, though invariably present, can very seldom be seen, since the background also is so strongly absorbed.

Some dyed turquoise has been found to give a broad absorption band centred at 655nm. This reminds us of the similar band found in stained jadeite, although the colours are quite different.

32 Absorption Spectra of Zircon

In Chapter 3 an historic letter from Professor A.H. Church was quoted, in which he described the remarkable absorption bands seen in zircon and almandine garnet. In the original articles that letter was quoted again at some length here. Such repetition is superfluous in book form, and it is only necessary to remember that Church was the first person to suggest that absorption spectra might be used to identify minerals.

Two years after the Church letter, the zircon bands were independently observed by the well-known mineralogist, H.C. Sorby, who ascribed their presence to a new element which he christened 'jargonium'. Sorby, unaware of Church's article, published his results in the Proceedings of the Royal Society,[1] and accompanied his paper with drawings showing as many as fourteen bands and depicting the effect as seen in polarized light corresponding to the extraordinary and to the ordinary rays. Later, Sorby was able to correct his first assumptions concerning the mythical 'jargonium' and to prove that the bands could be accounted for by the presence in most zircons of uranous oxide. If a trace of a uranium compound is added to microcosmic salt (hydrogen ammonium sodium phosphate) and fused into a bead in a loop of platinum wire, taking care to keep the fused bead in the reducing part of the blowpipe flame, a transparent globule is formed in which narrow absorption bands can be seen which resemble those in zircon to a marked degree, though differing in detail. Such a spectrum is shown in Fig. 32.1.

Another experiment by which a zircon-like spectrum can be induced artificially is to reduce an acid solution of a uranium salt, such as uranium nitrate, by adding pieces of metallic zinc. The zinc dissolves in the acid, emitting nascent hydrogen, which reduces the uranium to the uranous (divalent) condition. As the reaction proceeds the observer can detect the gradual emergence of the 'zircon-like' bands which replace the vague, evenly spaced bands typical of uranyl uranium. In such experiments the uranium is of course present in major quantity, but Sorby was able to show that the presence of zirconium greatly enhanced the intensity of the bands, and less than 1 per cent of uranium in the mineral zircon is sufficient to yield absorption bands of remarkable intensity.

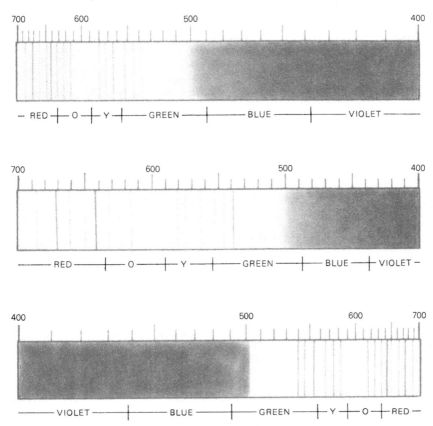

Fig. 32.1 URANYL CHLORIDE. A spectrum illustrating the similarity of the known compound of uranium with the zircon spectra

The strength of the zircon spectrum varies considerably from specimen to specimen – and although locality rather than colour seems to govern its intensity, it can in general be said that *red* zircons show the uranium bands least of all, and there are even some stones in which no lines can be detected. Until fairly recent times practically all zircons used in jewellery stemmed from the gem gravels of Sri Lanka. In their natural state, these were generally shades of yellow, brown and green. Some of them were found to lose their colour when heated, and colourless zircons, some of which from the Matara area in the south were natural in colour, were formerly sold under the misnomer 'Matura' or 'Matara diamonds'.

Most Sri Lankan zircons show pronounced absorption patterns of up to eleven bands, but in many green stones these are much less sharply defined, and in some cases may consist of just one woolly band, centred at 653nm in the red, where the strongest and most persistent zircon line is seen.

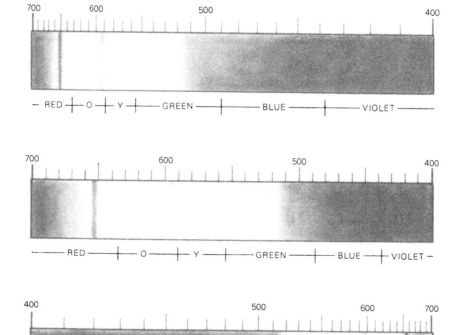

Fig. 32.2 LOW ZIRCON. With the crystal lattice largely destroyed the 653nm band remains as a tell-tale indication of zircon

Metamict green zircons

This modification of the normal zircon spectrum is due to the break-down of the crystal lattice under internal bombardment during hundreds of millions of years. The missiles in this bombard-

196

ment are alpha-particles (helium nuclei), which are emitted by the uranium atoms present as impurities in the mineral. If the refractive index and density of the green zircons showing only the one woolly band are measured, it will be found that they belong invariably to the 'low' type of zircon, with a refractive index near 1.80 and a density near 4.0. The scientific name for this is 'metamict'. A metamict mineral is one in which the chemical composition and outward crystal shape is that of a known crystalline mineral, but in which the crystal lattice has broken down due to radioactive bombardment, so that the physical properties are no longer the same and the substance is now practically amorphous. Metamict green zircons gradually recrystallize when heated, though in the case of very low types such heating must be intense and prolonged. In the case of stones in which the break-down is only partial, it is interesting to note how much crisper the absorption bands become after even a short heating to redness in the flame of a Bunsen burner. But in most cases the colour is by no means improved by this, so it is as well to limit the experiment to expendable fragments of rough green zircon. It is perhaps interesting that this rare metamict condition seems so far to be peculiar to the immensely old minerals of the Sri Lankan gem gravels, and to date only one other mineral has been found to display the same broken-down characteristics – the rare gem ekanite, which also comes from Sri Lanka.

Burmese zircons

Although Sri Lankan stones often show a strong absorption spectrum, the greenish-brown zircons found in Burma display the same bands in far greater strength, and in addition a number of extra bands can be detected, very fine and very faint between the giant pillars which form the main spectrum. Sorby, a micro-geologist 1826–1908, was able to depict fourteen bands in the spectrum of a Sri Lankan stone: in some Burma specimens we have been able to account for no fewer than forty-one bands and lines – an incomparably richer spectrum than that shown by any other gemstone.

Heat-treated zircons

Since the 1930s almost all zircons used in new jewellery have come, not from Sri Lanka or Burma but from localities in the region which was once called Indo-China, principally the countries of

Thailand and Cambodia, although the latter source was seriously disrupted by war and political strife for many years and gem mining almost completely ceased. Most of the stones were reddish-brown in their natural state, but were roasted in primitive ovens under controlled, but rule of thumb conditions, to produce sky-blue and greenish-blue, colourless and golden-yellow stones which became commercially very popular. Colourless stones had been seen before, but the two coloured varieties were startlingly different from those which had formerly been available from Sri Lanka. In the untreated stones it is difficult to detect any absorption lines, but once heated they almost invariably have the strongest uranium line at 653.5nm and a weaker line at 660nm. These are fine and

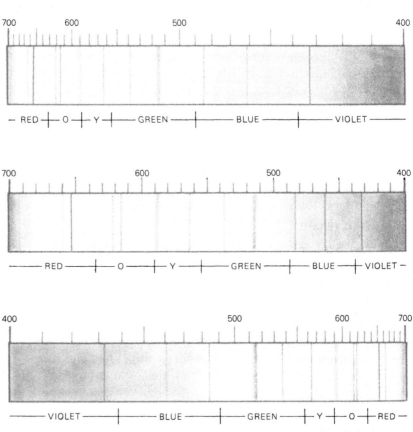

Fig. 32.3 SRI LANKAN ZIRCON. The spectrum varies in strength and in width of lines but is instantly recognizable for its ladder-like structure. This one shows twelve lines but the number and thickness can vary

198

usually faint, but are very distinctive, and are complete proof that the stone is indeed a zircon. If the student finds difficulty in seeing these by transmitted light the reflected light method may help, since the long path into and out of the stone enhances the absorption, and some other zircon lines may make their appearance.

Below is a list of all the bands and lines we have been able to measure in zircon, including some only seen on a photograph taken in a small grating spectrograph using special photographic plates sensitive to the deep red. The stronger bands (which are those seen in Sri Lankan zircons) are printed in heavy type, and the letters o and e are added to those lines which are markedly stronger in either the ordinary or the extraordinary spectrum, respectively.

756nm	o	626.5	**537.5**	
714	e	**621**	532	
703	e	**615**	527	
695	o	595	522	
691		**589.5**	o	**516**
686.5		583	e	511.5
e **683**		577	494	
678.5		574.5	**484**	
662.5		566.5	475	
e **660.5**		**562.5**	**460**	
o **653.5**		558	**432.7**	
o 646		551	426	
o 641		547	420	
o 636	e	543		

These bands are mostly narrow, as absorption bands in solids go; but their width varies according to the intensity of the spectrum. The main band at 653.5nm, for instance, varies in width from 1 to 4 nanometers, while the sharpest and finest of the stronger bands is that at 562.5nm, which is only 0.8nm in width.

We have been at some pains to establish as accurately as possible the exact wavelength of the more important bands, using the method explained earlier, of comparison with bright lines emitted by certain elements in the spectral region close to that of the wanted band. The wavelength of these bright lines is accurately known, and obtainable from tables. To mention a few examples: the zircon band at 691 was checked against a potassium emission line at 691.1; the 653.5 band by comparison with the barium line at 652.7 and strontium 654.7nm; while the 589.5 band is conveniently checked against the well-known yellow doublet of sodium at 589 and 589.6nm.

Anomalous zircon spectra

Although, broadly speaking, the only differences between the spectra of zircons from different localities or of different colours is one of intensity, there are stones in which the spectrum is markedly different from the usual. Such aberrant spectra seem confined to certain low-type zircons. One of these happened to be the stone in which we first observed the zircon bands, and for this and for the less trivial reason that part of the stone was utilized by Professor Chudoba[2] in his valuable researches into the metamict structure of low zircons, may be called a 'historic' specimen. We retained one half of this stone in our collection. The stone has the exceptionally low density of 3.965 and refractive indices 1.792-1.796. It is dull green in colour. The spectrum is notable in showing three broad strong bands in the red, where most metamict zircons only show a single vague red band. The lowest band is the usual 653.5 band: to the long-wave side of this is an even stronger band at 669nm, about 10nm broad, while the upper of the three bands is really a doublet consisting of the 691 and 686.5 bands of the normal spectrum. Of the remaining bands, some coincide with those of the normal spectrum, while others are anomalous. A drawing of this spectrum is given in Fig. 32.5. Such specimens are rare and we have seen only very few among the thousands of zircons that have passed through our hands which have had this 'three-band' absorption in the red.

Of about equal rarity is another strange variation in the spectrum of low zircon. In this, the usual intense 653.5 line is represented by a broad, vague band centred at 655, and there is a rather strong band at 520, and sometimes another band at 456nm. It is curious, and possibly significant, that each of the three or four stones of this type which we have encountered has had a very similar refractive index and density (1.82 and 3.89 respectively), with no birefringence.

Summary

Summing up the whole subject from a practical point of view, it can be said that almost all the zircons used in jewellery show a sufficiently distinctive spectrum to enable their identity to be established very quickly by this means alone, whether the stones are mounted or unmounted, large or small. The spectroscope can also quickly pick out rough zircon pebbles such as are found abundantly in the gem gravels of Sri Lanka. The heat-treated blue, colourless or golden zircons from Indo-China may show only the 653.5 band as a narrow line, accompanied by the fainter 660.5 band. These might easily be missed if one did not know where to look for them, and the beginner may have to use reflected light to see them clearly. In nearly all the natural-coloured Sri Lankan stones, there is a stronger and richer spectrum, with a series of strong, fairly evenly spaced bands striding down into the violet, while the rarely encountered Burma zircons are recognizable at once for their still richer spectrum.

Drawings of typical Sri Lankan and Burma zircon spectra are reproduced in Figs. 32.2, 3 and 4.

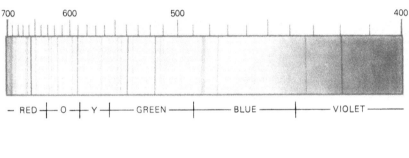

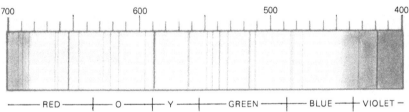

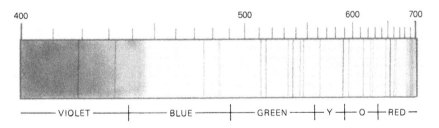

Fig. 32.4 BURMA ZIRCON. The strongest zircon spectrum which can give up to 41 lines of considerable intensity. This one has 23

In a previous article on the absorption spectrum of zircon,[3] we published photographs which showed very well the increasing strength but essential similarity of the bands in zircons from the three main localities. Dr Herbert Smith used this same photograph to illustrate the 1940 edition of his great book *Gemstones*.

Under ultra-violet light or when stimulated by X-rays, zircons often show a yellow glow, and this, when examined with the spectroscope reveals narrow bright lines in the yellow and green. This is due, not to the uranium content, but to traces of rare earths. This, however, is not the appropriate place to deal with this interesting effect in detail.

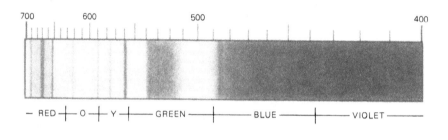

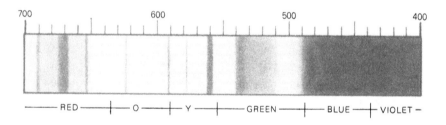

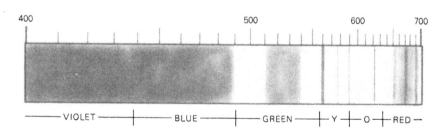

Fig. 32.5 ANOMALOUS ZIRCON. A rare 'low' form in which the 653 line has two more lines on its red side to make a prominent group of three. Other lines may not coincide with the normal zircon spectrum. The effect seems to be due to heating after millennia of metamictization

33 The Rare Earth Elements

The group of closely related elements known collectively as the 'rare earths' constituted one of the major research problems for chemists of the later nineteenth century.[1]

These metals are now conveniently divided into three groups, in accordance with differences in the properties of their compounds. These are, the Cerium Group, comprising lanthanum, cerium, praseodymium, neodymium and samarium; the Terbium Group, which includes europium, gadolinium and terbium; and the Yttrium Group, consisting of dysprosium, holmium, yttrium, erbium, thulium, ytterbium and lutecium. There remains the element of atomic number 61, which was eventually named promethium (Pm), but which has yet to be isolated from any natural source. It was found in atomic fission residues in 1945, when only two grams were isolated, and two kilograms were obtained later at Harford. It closely resembles neodymium, but little is known of its compounds.

Within each group, compounds of these elements are so similar in behaviour that it was almost impossible to separate them from one another by chemical means. Fractional decomposition and fractional crystallization of their salts were among the laborious methods used to obtain more or less pure compounds of the metals. Even in the periodic classification of the elements the rare earths still caused trouble, as there seemed no simple scheme by which they could be made to take their place in Mendeleeff's classical table. For a long while even their valencies were in doubt.

Electron distribution

It is now recognized that the chemical behaviour of atoms is chiefly dependent upon the nature of the outer electron shells, since it is by the sharing, giving or taking of these outer electrons that the elements can combine to form chemical compounds.

When the electronic structure of the rare earths is considered, we are provided at once with some explanation of the remarkable similarity of their chemical behaviour, since the transition from one element to the next here involves no change in the outer electron shell, but instead, a steady 'filling up' of a deeper lying electron group.

The electron orbits of the elements are classified from the innermost series outwards, into groups to which are allotted the letters K, L, M, N, O, P ... All the rare earth elements contain the standard number of electrons, 2, 8, 18, for the completed K, L and M orbits. For the N, O, P orbits the number of electrons are as follows:

Lanthanum	18, 8, 3
Cerium	19, 8, 3
Praseodymium	20, 8, 3
Neodymium	21, 8, 3
Promethium	22, 8, 3
Samarium	23, 8, 3

and so on, up to Lutecium, which has the grouping 32, 8, 3. The N electron shell has now been completely filled, and the next element, hafnium, which is not a rare earth, sees the resumption of the building up of the outer electron shell, with 32, 8, 4 for the N, O, P electron orbits.

This matter of electron orbits has been dealt with in some detail, as only thus can we explain not only the extreme chemical similarity of these rare elements but also the peculiarities of their absorption and fluorescence spectra. Both the fineness of the absorption lines which compounds of several of these elements display, and the fact that the wavelength of these lines is very little affected by the nature of the host crystal (or by the solvent in the case of solutions) are explained by the fact that the outer electrons are not involved in these absorption processes, but act as a screen behind which the electron shifts in the inner (N) orbits, which cause absorption or fluorescence in these elements, can take place relatively undisturbed by the surrounding conditions.

Spectra due to didymium

Many of the rare earth compounds, in particular those of 'didymium', erbium, europium and samarium, show well-defined and narrow absorption lines in the visible and near ultra-violet spectrum. Of these, gemmologists are concerned almost exclusively with the absorption bands due to 'didymium', which has been placed in inverted commas (from which it can now be released) to indicate that it is not, as originally thought, a single element, but consists of a mixture of the two almost inseparable metals, neodymium and praseodymium. In nature they always occur together and each contributes its own lines to the combined spectrum, which in practice is always referred to as the didymium spectrum.

The didymium spectrum is highly characteristic, and, since the nature of its 'host' causes very little variation in wavelength or character, it can be recognized with confidence whenever it occurs. The main pattern of this spectrum is a group of fine lines in the yellow region. As usually seen in a hand spectroscope, the lines are five in number, and so closely spaced that, unless the spectroscope is carefully focused and the slit narrowed, they may appear as a single broad band.

In yellow and green apatite in which, of all gem minerals, these appear most consistently, the lines of the yellow group are at 597.5, 585.5, 582.5, 577.2 and 574.2nm. There is a further group of lines in the green although these are much less intense. These are at 533.5, 529.5, 526.5, 525 and 521nm.

The yellow group occurs in greater or lesser strength in almost all colours of apatite and has been seen in brown, pale pink, pale blue and purple stones in addition to the classic yellow variety from Durango and various greens. Some of these will also show the green group of lines faintly. Deep blues from Brazil absorb all red, orange and green wavelengths and no didymium spectrum is to be seen. A pale blue variety from Burma, which often gives cat's-eye stones, sometimes has a different absorption which will be described below.

Probably the commonest example of a didymium spectrum is provided by the glass containing these rare earths which is sold under the name of Crookes glass (after Sir William Crookes, the famous scientist) for use in anti-glare spectacles. The reason why this glass is anti-glare is in fact due to the strong absorption in that part of the spectrum to which the eye is most sensitive. The colour of these lenses is usually a very pale pink. Students wearing Crookes lenses have been puzzled by the persistence of an absorption band in the yellow which they see in all spectra they examine. Among pastes a deep red stone and a very pleasing brown one also showed very strong didymium spectra, but these may have contained an extra chromophore element. As might be expected in an amorphous host, the spectrum in these glasses is not as sharp as when the lines are seen in crystals – but the difference here is less than it would be in the case of colorant elements other than rare earths.

Normally it is only apatite among the regular gem minerals that exhibits the didymium lines at all strongly, but a number of other minerals may be found which sometimes show vestigial traces of the yellow complex in their deeper coloured specimens. These include danburite, sphene, fluorspar, idocrase and calcite, in all of which the lines are usually very faint and elusive, and often totally absent; and scheelite, which in its orange variety may rival or surpass apatite in the strength of its absorption spectrum. Consideration will show that all these substances have in common the fact that they are essentially *calcium* minerals. There can be no doubt that chemically and physically, didymium behaves so similarly to calcium that small quantities of the rarer element have a tendency to follow it around, so to speak, when minerals are crystallizing from magma or being formed by secondary processes. Even the very rarely cut gems strontianite and wulfenite, which can contain calcium as a replacement element, have occasionally been found with a didymium spectrum.

While, therefore, the detection of didymium lines is not sufficient in itself to identify a gem (except in apatite where the strength of the lines is so outstanding that there can be little doubt as to its identity) it does give an indication that one is dealing with a calcium mineral, which may serve as a useful confirmatory test.

The absorption spectrum of didymium in YAG and the lines seen in yellow apatite are shown in Figs. 33.1 and 2.[2]

Blue apatite spectrum

The didymium spectrum of yellow apatite has been fully described and illustrated, and other colours of apatite have also been found to yield quite a strong spectrum of the same kind. It is surprising, therefore, to find that some attractive blue apatites which are found as slightly worn crystals in the Mogok district of Burma show no trace of these didymium lines but, instead, a quite strong series of narrowish bands of a different type and wavelength.[3] The origin of these bands is not so far known to us, but from their fair degree of narrowness they might well be due to some rare earth element other than didymium. Whatever their origin may be, this seems an appropriate place to describe this interesting spectrum. As many readers will know, the dichroism of most apatites is quite feeble. But in these blue crystals from

Burma it is a remarkably strong feature. Unfortunately it is the extraordinary ray which has the strongest colour (a truly magnificent blue) which means that, unless the artificial aid of polaroid is called in, the gem, even in its most favoured directions (at right angles to the optic axis) can only show this fine blue at half strength, diluted as it must be by the pale yellowish tint of the ordinary ray.

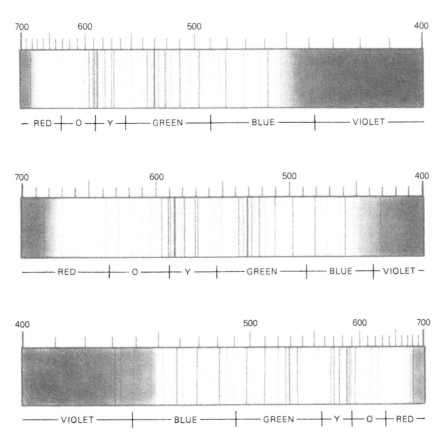

Fig 33.1 YAG DOPED WITH NEODYMIUM. A most beautiful absorption of about 44 lines obtained by doping the near perfect lattice of artificial yttrium aluminium garnet with Nd. This is the didymium spectrum par excellence

Yet it is in this pale ordinary ray that the absorption bands are seen, and if one observes with the spectroscope light which has passed along the axis of a crystal an inch or so in length, the

207

strength and clarity of the absorption bands are quite remarkable. There is a band of moderate strength in the orange, at 631, accompanied by a weaker band at 622nm. A very weak and vague band appears at 525, then a strong and narrow one at 511nm at the end of the green. A very weak band near 507 is followed by a strong, broader band at 490nm. Finally a vague, weak, and broad band can be seen in the blue, centred near 464nm.

In passing, it may be mentioned that the density of this blue apatite is 3.184 compared with 3.213 for yellow apatite from Durango, Mexico, which suggests some differences in composition – though the refractive indices, 1.634-1.636, are much the same in each case.

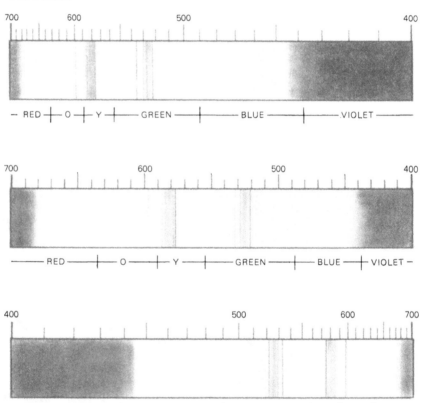

Fig. 33.2 YELLOW AND GREEN APATITE. A fine example of didymium in a natural mineral which may be found in varying strength in other colours of apatite, and very faintly in some other calcium containing gems. The rare orange scheelite has the same absorption quite strongly

208

Didymium in emerald?

Another gem mineral in which didymium lines may occasionally be seen, and at great strength, is emerald from the Muzo mine in Colombia. Beryl is not a calcium mineral, and this might seem to contravene the generalization made above. But in this case the didymium lines are due not to absorption in the emerald itself, but to inclusions of *parisite*, a rare earth mineral of which the type locality happens to be this same Muzo mine.

A remarkable example of such an emerald from the collection of the late Sir James Walton was examined by the authors[4] and found to show a strong didymium spectrum only when the light was transmitted through a large yellowish inclusion near the base of the (cabochon) specimen. Narrow bands or lines were measured at 588, 583, 579, 574, 571, 521, 513, 483, 469 and 464nm. There could be no doubt that the inclusion was parisite, a rich repository of the cerium earths of which the formula can be written $(Ce, La, Di)_2 (CO_3)_3.CaF_2$. It is interesting to note that the mineral was first called musite from the Muso (Muzo) mine, but later altered to parisite (after J.J. Paris, proprietor of the mine) as the name musite had already been used for a type of amphibole.

The Gemological Institute of America have reported two examples of the parisite spectrum both of which, although basically recognizable at once as didymium inspired, vary considerably in intensity and in line content, as might be expected when the possible varied content of rare earths is taken into account.[5,6]

Maxixe beryl

Here we may deal with the strange deep blue beryl already mentioned briefly in Chapter 27 of this book. This was found in an alluvial deposit at the Maxixe Mine, south of Rio Arassuahy in the province of Minas Gerais, Brazil, and was first described by G.O. Wild in 1934, and later analysed by Schlossmacher and others.

This extraordinary blue beryl has a caesium content of 2.8 per cent, and this is reflected in a high density (2.79-2.80), and high refractive indices (w 1.592 e 1.584). Unlike most aquamarines the ordinary ray has the darker colour – a deep blue when the material is freshly mined – while the extraordinary ray is almost colourless. Unfortunately this variety fades on prolonged exposure to light, and eventually assumes a rather uninteresting yellowish tint. The ordinary ray, while deeply coloured, shows a series of strong, narrow absorption lines extending, in diminishing intensity, from the deep red into the yellow.

The spectrum resembles that of zircon to a certain extent, and it is thus possible that uranous uranium may again be the cause of the bands. We have made the following measurements on a rough specimen given us by Mr Wild: 695 v.s.; 675 v.v.w.; 655 s.; 628 v.w.; 615 m; 581nm v.w. An exceedingly weak and vague band near 550nm was also measured.

Irradiated beryl

In 1973 a very fine dark, almost sapphire blue beryl appeared on the market, without explanation or origin other than 'Brazil'. This material had absorption bands very similar to, although not identical with, those for natural-coloured Maxixe aquamarine. It also shared the unique feature of the latter in that the fine blue colour belonged to the ordinary ray, and so did not suffer the dilution which is a feature of all other aquamarines. But it also shared the Maxixe material's unfortunate ability to fade in a matter of days upon exposure to strong light. Material of this deep colour was eventually recognized to be the product of irradiation treatment. It is easily identified by its unusual deep blue, which cannot be made to look darker by using a polaroid filter; and by its very remarkable and unusual absorption spectrum, which is illustrated in Fig. 33.3.[7] These stones fluoresce greenish-white under short UV light, and give a red residual colour through the Chelsea emerald filter.

It is worth noting that a partially faded irradiated stone might possibly be mistaken on sight for a normal aquamarine of good colour, but the tests outlined above would reveal the deception at once.

Although offered originally at very high prices, these two variations of the aquamarine saga, whether Maxixe or irradiated, are best avoided because of the ephemeral nature of their colour, which can eventually fade to pale yellow, or pink or even become almost colourless. Re-irradiation may restore them, but the 'fade' is a built-in factor of the mineral structure and will repeat itself.

Green andalusite

The final spectrum to be described in this 'rare earth' section is that of a bright green, very dichroic form of andalusite found, it is believed, in Brazil, but the exact provenance of which is not known to us. Small waterworn pebbles of rough, and even smaller cut stones, have been encountered from time to time in old collections, and we possess a fair-sized polished piece which was given to us by the late Mr Edward Hopkins, on which the measurements which follow have been made.

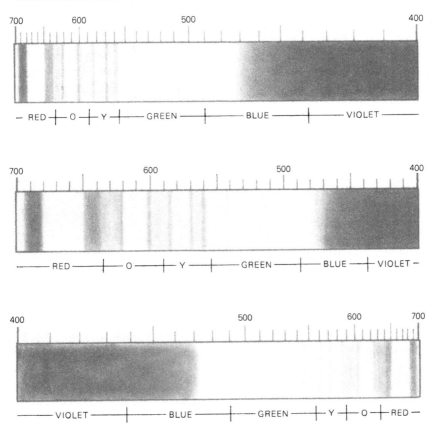

Fig. 33.3 DEEP BLUE AQUAMARINE. An almost sapphire blue form of beryl produced from pale material by artificial irradiation gives this extraordinary absorption in its ordinary ray only, which is the coloured one in this material. A natural gem from the Maxixe mine in Brazil has a similar spectrum, but both types fade quickly in strong light

Curiously enough, three small specimens of this form of andalusite once turned up during a routine test on a parcel of 117 green stones, of which 111 were demantoids.

We first observed the remarkable spectrum shown by this andalusite in 1933, and in the battered old notebook in which all our earlier observations were recorded the following remark is to be found. 'These lines in the green shown by our andalusites are as fine as, if not finer than, any absorption lines we have seen.' That observation still remains true.

The principal bands are in the green. The first grades in intensity, ending in a knife-edge at 552.5, followed by fine lines at 550.5, 547.5, with a second group at 518, 507, 500 and 495nm. Blue and violet are strongly absorbed but a band at 455 may be visible. This band has been noted in more normal types of gem andalusite, sometimes accompanied by a narrow band at 436nm. These bands in the blue and violet are not rare earth bands, and are probably due to iron.

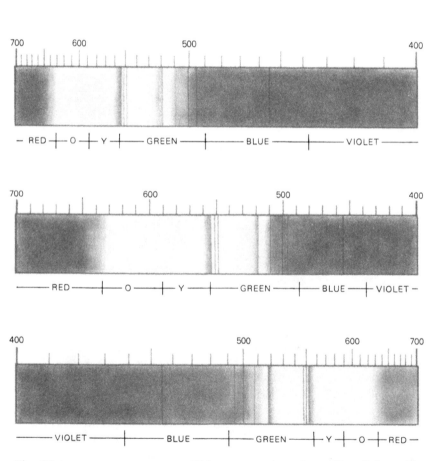

Fig. 33.4 GREEN ANDALUSITE. This rare variety from Brazil has this strange spectrum of bands with sharp edges to their blue sides. Probably due to a rare earth but which one is not known

While the sharpness of the green lines in andalusite strongly suggests a rare earth origin, it must be confessed that no lines in this region have been seen in any of the authentic rare earth spectra observed by ourselves or by Mr T.H. Smith (who specialized in such spectra). Since yttrium earths, in particular erbium, had been suggested as a likely cause, we have checked on the absorption bands in the mineral xenotime, which is an yttrium phosphate rich in erbium. A group of very narrow bands in the red near 650 were seen, and a doublet near 520nm in the green, in addition to vague bands in the blue. The position of these bands is strikingly similar to the (admittedly ill-defined) bands noted in a few anomalous specimens of 'low zircon'.

The delicate beauty of the green andalusite spectrum is difficult to describe in words, but the drawing, Fig. 33.4, should provide readers with a good idea of its appearance. Lines toward the blue end may be obscured by the general absorption which can extend up to about 500nm.

34 Absorption Spectra of Diamond

Diamond is by far the most important gemstone, and it may therefore seem strange that a description of its absorption spectrum should have been left until so late in the book. The reason is simple enough, for carbon does not happen to fall into any of the transition element categories (such as chromium, iron, etc.) which have so far been considered.

At the time when the original articles were written it was believed that such absorption bands as were seen in diamond were due, not to foreign elements in the stone, but to the diamond structure itself or to defects in that structure.

However in 1959, Kaiser and Bond of Bell Telephone found that Type I diamonds contained small amounts of nitrogen. Subsequent research by Dyer and others revealed that it was the large size of this atom when compared with that of carbon which was responsible for the defects in the diamond crystal structure. Further work by Dr A.T. Collins and others has shown that nitrogen can be present in one or more of several configurations. Type Ib diamonds have very little nitrogen and what there is, is present as single atoms disseminated through the crystal structure, each one remote on the atomic scale from any other nitrogen atom. Such stones absorb generally in the blue region without producing any obvious line spectrum, and are an attractive yellow in colour. They are quite rare.

Most diamonds have much more nitrogen, and heat and pressure over aeons of geological time have caused these nitrogen atoms to collect together to form well-defined clusters. Such diamonds are known as Type Ia and are then sub-divided into A, those in which the nitrogen atoms are in molecular pairs, very strongly bonded together, and B, in which the clumps are of larger collections of nitrogen molecules. A third type of aggregate is known as N3 and apparently consists of three nitrogen atoms around a single carbon atom.

Other more extensive structural defects, which can be large enough to be visible with an electron microscope, are now thought to contain comparatively little nitrogen.

All such aggregates cause marked absorption in the infra-red and ultra-violet regions of the spectrum but they do not, on their own, cause absorption in the visible wavelengths, so diamonds with a lot of nitrogen are usually colourless.

But most diamonds have mixtures of the different types of nitrogen structure, and the N3 aggregate occurs in conjunction with either or both of the other types. When this occurs in a stone with a high concentration of the large B type clusters there is an increase in the strength of the 'Cape' absorption bands and a fall-off in what the trade considers to be fine colour. Some rare diamonds which show no infra-red absorption contain practically no nitrogen and are known as Type IIa.

It has already been mentioned that when the temperature of a tested stone rises under the heat of a high-intensity lamp, the absorption lines become progressively more and more diffuse, due to the intrinsic oscillation/vibration of the crystal's atoms which, from theoretical stillness at 0°K, become more and more agitated as the temperature increases. This is true even at room temperature, for at 18°C the crystal is at about 290°K in absolute zero terms and the absorption pattern is already subject to considerable scattering.

In recent years the need to see elusive spectra in diamonds coloured by artificial irradiation has initiated a laboratory technique of concentrating the scattered absorption lines by lowering the temperature of the test stone as far as is conveniently possible.

Anderson made tentative steps in this direction by cooling the stones on 'dry ice', solid carbon dioxide. There were problems with stones which frosted over because they were super-cooled in the ordinary moisture-laden air of the room. But temperatures of around 200°K were probably reached, sharpening the spectra to some extent, but still rather too high to concentrate the lines as much as was desired.

Zero K is unobtainable, so a compromise is made by testing the stone in a Thermos-type, double-walled glass chamber, to exclude moist air, in a stream of cold gas evaporated from a large thermally-insulated flask of liquid nitrogen. This gives a working temperature of about 120°K at the stone, about 170°K below room temperature, so that the absorption spectra will be considerably sharpened. Work has been done with stones at even lower temperatures, but it is reported that this further reduction does not improve the spectra sufficiently to justify the considerable expense and difficulty in getting down to such extremes of cold.

It will have been noticed that we are using nitrogen in order to see absorption which is itself due to nitrogen atoms in the diamond. Gaseous nitrogen has no absorption in the regions we are examining and has no effect whatsoever on the absorption of the stone beyond making it sharper by cooling and thus reducing the vibration of component atoms within the crystal. So at these much lower temperatures absorption lines and bands become more obvious and other lines previously unseen may appear.

Having briefly explained the causes of colour in diamonds, and the methods used to obtain the best possible absorption spectra from them, we now need to describe the absorption of this important gem.

The first man to describe absorption bands in diamond was B. Walter (1891). His very careful experiments on some fifty assorted diamonds from different regions used sunlight as a light source, and estimated the wavelengths of bands seen by comparison with the Fraunhofer lines.

He noticed a narrow band at 4155Å (415.5nm) in the deep violet, and that it was more intense in yellowish ('Cape') stones than in colourless ones. When this one was strong, another band was seen at 471nm, while other faint bands were detected in the ultra-violet region in photographs of the spectrum.

Much later a classic paper by Robertson, Fox and Martin (1934) suggested erroneously that the 415.5nm band was not normal to diamond, since they had seen it in only one of the specimens they examined.

During the 1930s Anderson and Payne researched intensively into all gemstone absorption and came to agree broadly with Walter so far as diamonds of the 'Cape' series were concerned, and were able to detect further bands in the blue which he had missed, among them one at 478nm which appeared as the depth of colour of the diamonds became more marked (Fig. 34.1). [Here it

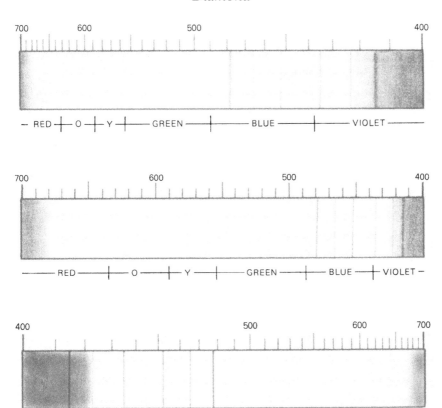

Fig. 34.1 CAPE DIAMOND. The typical absorption of yellowish 'off-colour' Cape stones. The line at 415.5nm is the key one

has to be remembered that they had the benefit of powerful artificial light sources, while Walter saw his absorption against the great lattice of the Fraunhofer spectrum which could quite easily mask the finer lines due to the diamond].

They also found that brown, and greenish diamonds of the same series, which fluoresced green, showed a completely different spectrum. Here the principal line is at 504nm, and this, as with the 415.5nm line in the normal series, could be seen as a fluorescent line under favourable conditions of stimulation. Other fine lines at 537 and 498nm were also found (Fig. 34.2), while diamonds of mixed green/blue fluorescence were found to show both the 504 and the 415.5nm bands.

217

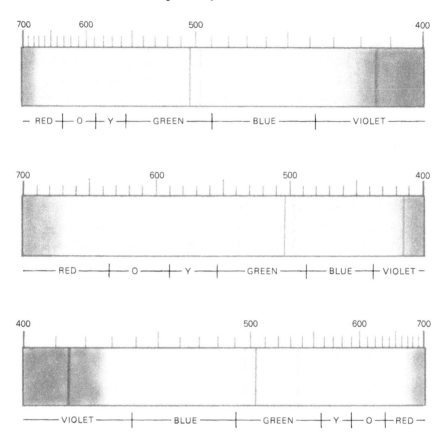

Fig. 34.2 BROWN AND GREEN DIAMOND. A different absorption pattern from that of the Cape stones, but the two types can be found in combination. The 504nm line is the key one

Under the leadership of Sir C.V. Raman, at the Indian Institute of Science at Bangalore, G.N. Nayar and Anna Mani thought that there was a direct connection between the intensity of the 415.5nm band and the strength of the fluorescence. Anderson and Payne were not able to confirm this connection but did establish a direct relationship between the depth of colour in 'Cape' stones and the strength of the 415.5 band, a factor which had not been considered by Professor Raman's team which was working mainly with colourless diamonds. Miss Mani did record one specimen with a 'distinctly yellowish tinge' as having 'anomalous' absorption bands in the blue. These were almost certainly the normal bands for a 'Cape' stone.

In the vast majority of cases diamond is recognized without recourse to spectroscopy and it is only when the vexed question of artificial colouring by irradiation rears its head that the spectroscope is needed to determine that fact.

Different types of irradiation produce different effects. Electron bombardment penetrates only about 2 mm. Gamma rays penetrate completely but take several months of exposure to achieve any marked results. Fast neutron bombardment also penetrates completely but causes enormous damage to the crystal structure. The alpha particles from radium again affect only the surface of the diamond, but leave a long-lasting residual radio-activity which has to be considered as a possible health risk.

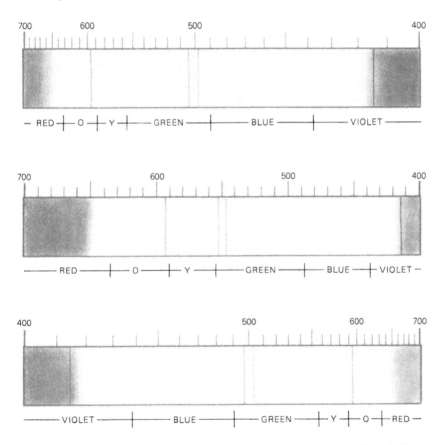

Fig. 34.3 IRRADIATED YELLOW DIAMOND. Shares the 'Cape' line, but 597, 504 and 497nm are the result of irradiation

Irradiation by whichever means usually results in a green stone, increasing to black if the treatment is continued. The diamonds are afterwards annealed by heat in order to lighten or to change them to a more acceptable colour. Most stones seem to be heated to temperatures below 800°C. Above that there is some risk of damage to the diamond by burning and, perhaps more importantly from a gemmological viewpoint, some telltale absorption lines are lost.

In the spectrum of a treated stone a line in the very deep red at 741nm is considered to be absolute proof of irradiation, but this is almost outside the range of human vision and needs photographic means to detect it with certainty. Another line or band appears at 594nm after annealing, but this disappears if the annealing temperature is allowed to go above 800°C (Fig. 34.3). Lines at 504 and 497nm also appear after annealing and are much more stable even when heated well above 1000°C. These may vary in their relative strengths, but both have also been seen occasionally in known natural yellow and yellow-brown stones. Dr A.T. Collins has said that the 497nm line is very rare in natural stones. When 504nm is strong, a phonon, or 'echo', may be seen at 496, giving the 497 line the appearance of a doublet.

Natural mauve and pink diamonds have a vague broad absorption band centred at 563nm (Fig. 34.4). The same band is to be seen in some natural brown stones. But irradiated pinks are made from the rare Type Ib yellow diamonds and, once treated, will give a sharp line at 637nm, and others at 595, 575 and 503nm may be seen (Fig. 34.5).

Natural blue diamonds are always Type IIb stones and are semi-conductors of electricity. Irradiated blues will have been made from other types of diamond and will not conduct electricity.

Some published spectral graphs may show certain 'key' lines in the diamond spectrum labelled rather confusingly by physicists. Thus lines at 741, 505, 497 and 415nm become known as GR1, H3, H4 and N3 respectively. These terms may then be used in the text without further reference to their actual wavelengths. This may be very convenient for the physicist, but it savours of obscurantism in texts read by the ordinary gemmologist.

Over the years many absorption lines in addition to the ones mentioned have been seen in diamonds of various colours, both natural and irradiated, and it is now evident that the whole subject of the absorption of this stone is complex to an extreme.

Almost all bands and lines in the visible light range now associated with the deliberate radiation treatment of diamonds have also been seen at one time or another in stones which are beyond doubt natural in colour, and we have to remind ourselves that such irradiation may have occurred as a result of chance mineral contacts at any time in the immense life history of individual stones.

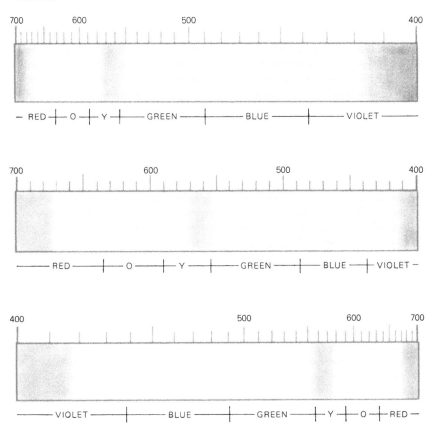

Fig. 34.4 NATURAL PINK DIAMOND. A vague broad absorption centred at 563nm

Recent research involving an assessment of residual radio-activity in garnets included in diamond crystals has put the age of the garnets, and that of the containing crystals, at more than three billion years, and a great deal can happen in such an unimaginable period of time! It also serves to underline the marketing slogan 'Diamonds are Forever!'

With some diamonds of fancy colour we may arrive at a situation where it is very difficult to say one way or another whether that colour is natural or artificially induced. Such stones may need to be submitted for official laboratory investigation when absorption patterns in the infrared region should put the matter beyond dispute. Such wavelengths are outside the scope of this book, and the tests involved beyond the reach of the ordinary gemmologist. But it is as well to know that they are available, even if rather costly. If the stones are of a significant size then such confirmation is vitally necessary since it will affect the value of the stone in question quite considerably.

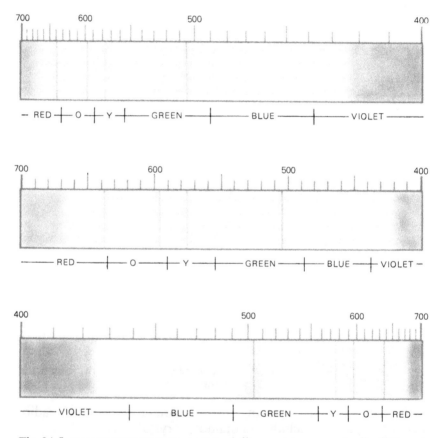

Fig. 34.5 IRRADIATED PINK DIAMOND. A line spectrum markedly different from that of the natural pink stone

35 Miscellaneous Spectra

Most of the important spectra found in precious stones have now been described and classified on the basis of the metallic oxide (colouring agent) which gives rise to their absorption lines or bands. But there are still a few which have not been dealt with, either because the cause of their absorption is not yet known, or because their comparative small importance as gems has led to their omission from earlier chapters. Our aim is to be as complete as possible, so these odd 'miscellaneous' spectra will now be dealt with.

Zinc blende, or sphalerite, although it is the most important ore of the metal zinc, is too soft a material to be satisfactory as a gemstone, though the rich yellow to orange-brown, red and even green of its transparent varieties, and its splendent lustre and dispersion give rise to stones of outstanding beauty, which unfortunately are all too easily scratched or cleaved and spoiled in wear.

Some, but by no means all, specimens of blende show narrow absorption bands at the red end of the spectrum, and these may cause confusion with zircons of similar colour and appearance. Thus a knowledge of the blende spectrum is useful for the collector of such rarely cut stones. The strongest and clearest of the blende bands is seen at 665nm. This is fairly narrow and in some specimens is a band of considerable intensity; in others it is weak or altogether absent. A much feebler band, also narrow, can often be detected at 651nm. Finally, there is a broader band of some strength in the deep red centred near 692nm. Owing to general absorption in this region (to which in any case the eye is very insensitive) this last mentioned band can scarcely be seen unless a good red filter be interposed to mask the glare from the yellow region of the transmitted spectrum.

The drawings (Fig. 35.1) will give a good idea of the fully developed spectrum. With brown and orange varieties there will be heavy absorption of blue and violet wavelengths.

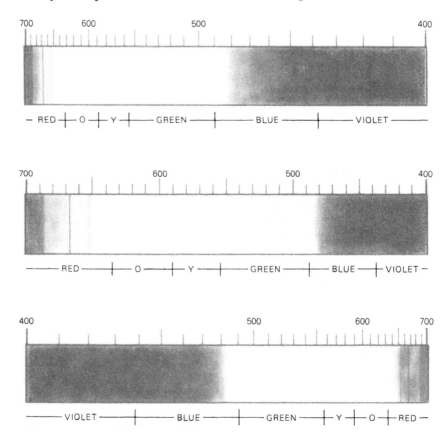

Fig. 35.1 ZINC BLENDE. Although too soft and easily cleaved to make a usable gem, blende is beautiful when skilfully cut, and has a spectrum reminiscent of low zircon, although its visual appearance and single refraction are quite different

The cause of these bands is obscure. Hartley and Ramage found by emission spectrum analysis, the presence of a rich variety of trace elements in blende – which is essentially a simple zinc sulphide. Elements detected included sodium, potassium, silver, copper, iron, gallium, lead, calcium, cadmium and indium. One or other of these impurities might be supposed to give rise to the absorption bands noted. Our own attempts to analyse by emission spectroscopy specimens of blende which showed the bands strongly for comparison with blende which did not show the bands, seemed to indicate a greater proportion of cadmium in the former. But that cadmium is indeed responsible for the absorption is by no means certain.

The *colour* of blende which in most mineral specimens is a very dark brown, is probably influenced chiefly by the iron content, although the positions of the spectrum bands would seem to deny this.

Absorption spectrum of sodalite

The blue mineral sodalite is not usually sufficiently translucent to enable its absorption characteristics to be easily observed. But in specimens sent to Mr Robert Webster by Mr D.S.M. Field, enough light could be transmitted to enable the spectrum to be studied. A strong and rather broad band was seen in the red, centred at 680nm, with a similar but weaker band at 595, and a very weak band at 540nm. These measurements are very approximate, as, owing to the feebleness of the light transmitted they could only be measured by means of a prism instrument with scale attached. The cause of the bands is not known. Maybe they are linked with the structure of the molecule itself, as in the case of the ultramarines, which are related substances.

Absorption spectrum of fibrolite

Fibrolite is potentially a gem of great beauty, though its easy cleavage makes the production of faceted stones a lapidary's nightmare. In a pebble, which was passed to us for identification by Mr Bernard Silver, we detected clear bands at 441 and 410nm, the latter being very strong. A weaker band was measured at 462nm. This specimen was found to have a density of 3.251, and the three refractive indices were measured as 1.6625, 1.6629, and 1.6837. Thus the specimen was undoubtedly fibrolite. The two stronger bands have since been detected in other specimens of the mineral, though far more weakly developed. It is possible that the absorption is due to ferric iron in this instance.

36   Absorption in Synthetic Gemstones

Although the absorption spectra of several of the colour varieties
of synthetic corundum and spinel have already been described or
mentioned under their appropriate sections, synthetic gemstones
play so important a role in determinative gemmology that it
should prove useful to have their absorption characteristics
discussed within the compass of a single chapter. This will also
provide an opportunity to include some information not previously mentioned.

Synthetic corundum

The principal colours in which the Verneuil corundums are manufactured are various shades of red and pink, blue, yellow, green,
brown, and that indescribable plum-coloured variety which is
intended to represent alexandrite. Colourless synthetic sapphire
has been very widely used in imitation of diamond, but it displays
no absorption bands and will not concern us here.

The reds and pinks owe their colour almost entirely to
chromium, and vary in shade from a deep garnet red to a pink
designed to resemble topaz. The vast preponderance of stones
have a bright and rather garish red colour approximating to that
of fine Burma ruby, while lacking the latter's richness and subtlety.
The spectroscope can give little help here in distinguishing
between these stones and natural corundums of similar depth of
colour. It is true that the absorption bands are a little more
emphatic (they make excellent demonstration stones) and the
fluorescence spectra a little brighter; but no great reliance can be
placed on such qualitative differences, and, fortunately, there are
many other more positive tests.

In recent years Chatham, Kashan, Ramaura, Knischka and
others have produced a variety of synthetic rubies made by the flux-

melt process, but once again the spectra of these stones in the visible wavelengths are only slightly different in emphasis from that of fine Burma ruby. Some workers have found significant differences when the ultra-violet region is examined photographically, but this is a test which is outside the scope of normal gemmology.

Synthetic blue corundum

It is a different story when we come to blue corundum, in which the spectroscope has proved itself of enormous value in distinguishing between natural and synthetic stones. It is well-known, since its discovery in our laboratory, that virtually all natural blue sapphires display a rather narrow absorption band, centred at 450nm, which is the strongest member of a three-fold absorption complex belonging in its full strength to green sapphire and attributed confidently to the presence of ferric iron. This band has been detected in a deep blue boule, in which the iron added with titanium to the raw material had obviously not evaporated entirely from the surface. But it is still true to say that the 450 band has yet to be detected in any *cut* synthetic sapphire – and the number tested by spectroscope runs, by now, into many thousands, in a wide variety of sizes and shades. Thus, where the band is clearly seen, it may be taken as conclusive evidence that the stone is a natural sapphire.

Absorption bands in synthetic sapphire are relatively so vague and feeble that they have hitherto been disregarded; but bands of a kind do occur in some specimens of this material, and are not too difficult for the practised spectroscopist to observe and make use of, when he knows just what to look for. This is a matter of considerable importance, since it means that the spectroscope can provide a positive instead of a merely negative test for synthetic sapphire.

E.T. Wherry, in his comprehensive paper, referred confidently to a band at 4780Å (478nm) which, he maintained, occurs more strongly in synthetic than in natural sapphires. This is something of a mystery, for we have never seen such a band in any sapphire-like stone, other than synthetic blue spinel.

The most noticeable of the weak bands in synthetic sapphire we wish to draw attention to is a vague smudge in the blue centred at about 455nm. It is obvious that this is uncomfortably near the 450 band of natural sapphire and it would need accurate measurement to distinguish between them on the score of wavelength. But it has always been one of our tenets, in advocating the value of the spectroscope in distinguishing between one stone and another, that, for diagnostic purposes, wavelength measurements are really unnecessary when one is a practised worker. All that is needed is that one should take into account the appearance of the stone, the general distribution and position of any absorption bands: the 'pattern' they make together, and their *nature* – that is, their breadth, strength, sharpness or diffuseness, complexity, and so on. This tenet still holds in the present instance: whereas the 450 band is quite narrow and well defined, the 455 band in synthetics is a vague blur of considerable width, and without any semblance of definition. Also, it may profitably be remembered that the 450 band in natural sapphire is stronger in the ordinary ray, so that by placing a polaroid disc over the eyepiece (or below the slit) of the spectroscope and turning this into the optimum position, a noticeable increase in strength may be observed, which may help matters considerably when the band is faint, as in Sri Lankan sapphires.

There is more to the synthetic sapphire spectrum than just this one blue band. On either side still weaker absorption bands may be detected or 'sensed'. One of these (the easier to see, since it is in a region of greater visual acuity) forms a vague blur near 490nm, where the green region has just given way to the blue. The other is a weak band in the violet near 428nm. The total effect is a very vague three-band absorption, and with practice the general appearance produced, weak though it is, can become distinctive, and certainly of more value than the mere absence of the 450 band, hitherto accepted as indicative of a synthetic stone, could be, since positive indications, however slight, are always more helpful than merely negative ones.

It is wise to give oneself every aid in attempting to observe such 'difficult' bands, and a blue filter will help considerably. But it is also necessary to remember that many blue synthetics do not show them.

In deep blue sapphires, whether natural or synthetic, a broad band can be seen in the green, centred near 560nm. This belongs to the 'ordinary' ray, and can be enhanced by turning the stone until the deepest blue ray is transmitted, or by interposing and turning a polaroid disc. This band can be quite strong, but it has very little diagnostic importance.

Synthetic yellow sapphires

These also show signs of a weak 455 band, and in the brownish-yellow types a strong absorption can be noted from this point onwards. Natural Sri Lankan yellows show either no bands or the 450, narrow and faint. Yellow sapphires from Australia, Thailand and Montana all contain enough iron to display the 450 band strongly, with the 460 and 471nm bands of the complex much weaker.

Synthetic green sapphires do not show a distinctive spectrum, but the purplish 'alexandrite' type invariably shows a single narrow line at 475nm, due to vanadium. This line, the strongly developed curved striae, and its distinctive colour, changing to raspberry red in artificial (tungsten filament) light, make this variety perhaps the easiest of all synthetic corundums to identify.

Chromium has been used in some of these stones to enhance the change to red, and will give a fluorescent line at 693nm.

Synthetic spinel

The most popular colours in synthetic spinels made by the Verneuil process are the various shades of blue, ranging from pale 'aqua-marine' tints, through medium 'zircon' blues, to the deep Reckitt's blue with red reflections that is sometimes used to represent sapphire. All these owe their colour to cobalt, and show the distinctive series of bands we have learned to associate with that element, when in a suitable state of combination. In deep-coloured stones the three strongest bands coalesce to form a solid block extending from about 670 to 510nm with weaker bands at 478 and 452nm. In paler pieces the three bands are seen clearly separated, with centres at 635, 580 and 540nm. In all cases deep red light is freely transmitted, and this, together with the absorption in the 580 region, causes the stones to show some shade of red through the Chelsea filter – forming a useful rapid check.

The synthetic sintered spinel designed to simulate lapis-lazuli which has been marketed in Germany, is also coloured by cobalt. This material is only faintly translucent, but a powerful beam of light allows a curious purple-coloured glow to filter through, which the spectroscope analyses into a narrow transmission band in the deep red, around 700nm, together with a feeble continuous transmission of the blue and violet. By reflected light a more normal cobalt spectrum can be recognized, with bands centred near 650, 580, 535, 480 and 455nm.

Many blue synthetic spinels (but not the 'lapis' type) show a red glow under ultra-violet light, or between crossed filters. This glow can be analysed into a narrow fluorescent line closely resembling that shown by ruby, but having a wavelength of 687 as against the

693.5nm of corundum. Feeble unresolved fluorescent lines, close to the short-wave side of the main one, can also be discerned. The 'organ-pipe' series of fluorescence lines which are so characteristic a feature of natural red and pink spinels are never seen in the Verneuil synthetics – which makes the 'organ-pipe' structure a valuable diagnostic test for natural spinel.

Fluorescent green spinels

Other synthetic spinels which show a distinctive spectrum include those of yellowish-green tint, which is due to manganese. These display a strong, narrow band in the violet at 423, with a broader and far weaker band centred at 448nm, where the blue merges into the violet. Faint bands due to cobalt may also be present. Some blue synthetic spinels also contain a little manganese, added to modify the colour, and these will also show the bands just mentioned, though rather weakly. The green synthetic spinels activated by manganese display a striking bright green fluorescence under long-wave ultra-violet light, which is also due to the manganese. An analogous instance is provided by specimens of the natural mineral willemite from Franklin Furnace, New Jersey, which has a brilliant green fluorescence for the same reason.

Synthetic pink spinels, coloured by iron, are not often encountered, but form quite attractive gems having a colour not unlike that of morganite beryl. In these we have observed two bands in the green which do not seem to have been recorded. The stronger of these is centred at 552nm, and the other band, which is very feeble, has a wavelength of approximately 580nm. Both bands are rather broad and ill-defined. In the case of feeble bands of this nature, in stones of pale tint, it may be found advantageous to make observations by light reflected from within the stone rather than to transmit light directly through them, as there is relatively greater absorption in the former method.

Synthetic emerald

When the first specimen of 'Igmerald', the German synthetic emerald, reached us in March 1935, we were very gratified to notice, when we examined its absorption spectrum, that two distinct bands could be detected at 606 and 594nm, in the extraordinary ray, whereas in all the natural emeralds we had hitherto seen there was merely general absorption in this region. This seemed to promise a useful means of distinction. As with synthetic ruby, however, it is merely a matter of how great is the chromium content of the stone. Synthetic emeralds do tend to be very chrome-rich, and to show an intense spectrum, even to the extent of showing these extra bands, but they *can* sometimes occur in exceptionally dark specimens of natural Colombian emerald, and it is therefore wiser to rely more on the evidence of inclusions and of the physical constants when seeking conclusive proof whether an emerald is natural or synthetic.

Synthetic rutile

Of the synthetic stones made by the Verneuil process, the yellowish rutile was found to show a powerful absorption band at about 423nm. Beyond this region there is practically no transmission, which partly accounts for its amazing dispersion – since refractive index always climbs steeply for wavelengths approaching a region of intense fundamental absorption.

We have not been able to detect any definite bands in colourless YAG, strontium titanate or cubic zirconia, but their coloured forms may often give marked absorption patterns. It has also to be remembered that with all synthetics, new colours and new methods of obtaining old ones may be introduced at any time, and will possibly give new absorption spectra. In this respect the synthetic is more flexible than the natural mineral.

Synthetic scheelite

Synthetic scheelite has been doped to give a number of different colours, some of which have rich absorption spectra, varying around an intense version of that due to didymium, which happens to be the spectrum seen in natural orange scheelite. So the spectrum does not help much in this case in deciding whether a stone is natural or synthetic.

The synthetic material is produced for laser purposes, and cut gems are unlikely to be seen on the general market.

Yttrium aluminium garnet

Yttrium aluminium garnet, or YAG, is a man-made crystalline material with a garnet structure, despite the fact that it contains no silicon. Colourless and transparent when pure, this has been doped with a number of transition and rare earth elements in the search for combinations which will 'lase' successfully in various new wavelengths.

Besides the fluorescent wavelengths needed for this purpose, many of the dopants give rise to absorption spectra which can be extraordinarily prolific in sharp absorption lines, far sharper than one would expect to find in any natural stone. Perhaps the nearest approach to such patterns is that found in well-developed zircons from Burma.

Different dopants result in markedly different absorption spectra, but one resulting from neodymium has already been illustrated in Fig. 33.1. Other dopants will obviously give different patterns but, when they occur, there seems to be an overall similarity in the sharpness and abundance of lines which is usually recognizable.

Such materials are often made in colours which are unlike the familiar ones of natural gems, and in many cases the material refracts above the normal limit of the standard refractometer, and may be detected only by using either an infra-red reflectometer or a thermal probe meter. These may not necessarily identify the substance with certainty.

These differences in absorption and colour, due to the use of different dopants or combinations of dopants, serve to underline one basic fact of many synthetics, that they are usually capable of being varied more or less at the whim of the manufacturer and, with the possible exception of the better-known Verneuil synthetics, cannot be relied upon to give constant testing factors through all possible specimens. Such variability applies equally to their absorption.

The variety of absorption patterns is to be expected more in these 'laser' synthetics, which are not produced primarily to supply the gem market, but are the result of experiments to find new laser materials. So all suitable dopants will have been tried and different absorptions will have been introduced with each new colour, many of which may show sharp absorption patterns in the visible region of the spectrum. Off-cut material with gem-like qualities of colour and transparency has often been sold and may find its way to the amateur cutter and thence to the gem market, usually being snapped up by collectors of gems rather than used in mounted jewellery. In a few exceptional cases the material will be produced with the gem market in mind and may then appear in quantity.

Another factor of the synthetic market which is not generally realized, is that each 'improved' synthetic tends to oust the earlier ones. Thus the ultimate appearance of cubic zirconia as a simulant of diamond which is almost exact in appearance, has largely removed the earlier imitants, synthetic white sapphire, synthetic white spinel, synthetic rutile, strontium titanate, YAG and GGG from the current gem market, although they will still be found in older jewellery. Flux-melt synthetics of ruby or sapphire are more convincing than the Verneuil equivalents, but the latter provide an exception to the rule just stated, in that their far cheaper production process keeps them in demand despite the improvements of the newer flux and hydrothermal methods of synthesis.

37   Absorption Spectra of Glasses

In contrast to crystalline gemstones, glass imitation gems seldom show any distinctive absorption bands. Even when bands are to be seen they are not clearly defined. In this, glasses are revealing their true nature as supercooled liquids, in which the viscosity is so high as to make the process of crystallization extremely slow – though, as a fact, unless the composition and thermal treatment of glass melts is carefully controlled, 'stones' consisting of small crystals or clusters of crystals do readily occur to mar the desired transparency of the product.

Of the elements which give rise to visible absorption centres in glass imitation gems, we can identify cobalt, chromium, didymium, selenium and uranium, with tolerable certainty. Copper, manganese, titanium, nickel and iron do not seem to produce definite bands.

Cobalt glass

The majority of the blue glasses intended to imitate sapphire are coloured with cobalt. Such glass is also used as the base of blue garnet topped doublets. The practised eye learns to recognize the peculiar reddish glints from these cobalt-blue counterfeits, and a first suspicion is quickly confirmed when they are viewed through a Chelsea filter, as they appear rich red in colour. The cobalt spectrum readily accounts for this: red light is freely transmitted; while the yellow-green region (which is the only other light passed by the filter) is heavily absorbed by the middle member of the three main absorption bands.

Synthetic blue spinel, of course, is also coloured by cobalt, and presents the same appearance through the Chelsea filter and a very similar series of absorption bands. The position of the bands in glass is rather different from those in synthetic spinel, the distance between the two outer bands being greater. But the safest guide, and one needing no measuring device, is to note the relative width of the central (yellow) band compared with the third band (in the green). In glass the central band is distinctly the narrower (in appearance this is enhanced in a prism spectroscope, where the 'spread' becomes greater towards the shorter-wave parts of the spectrum) while in the synthetic blue spinel the central band is markedly broader than the band in the green. Typical measurements are as follows, the width of each band being given in parenthesis:

| Cobalt glass | 656 (58) | 589 (22) | 534 (25) |
| Synthetic spinel | 633 (43) | 580 (40) | 541 (13) |

There is another band in the complete cobalt spectrum which is seen as a vague blur centred at 490nm in glass, and more clearly cut, though weak, at 480nm in synthetic blue spinel. These two spectra have been dealt with in greater detail in Chapter 30, and illustrated in Figs. 30.2 and 30.3.

All blue materials, or greens in which cobalt is involved as a colorant, transmit freely in the red end of the spectrum, and this makes them dangerous for signalling purposes, since distant lights shining through such filters may appear red to the observer, especially when mist or fog intervenes, through which red light penetrates more freely than light of shorter wavelengths. For green signal glasses copper is the ideal colorant, as every trace of red light is filtered out from the original bright source by this element.

Chromium in glass

When chromic oxide is dissolved in glass, its sharp line spectrum disappears and an absorption resembling that shown by an aqueous solution of some chromic salts is all that remains. In some of the green glasses cut as imitation gems we have been able to measure vague bands in the red. The strongest was centred near 682nm, and two vaguer and weaker bands appeared at 661 and 634nm. A microcosmic salt bead to which traces of chromic oxide were added became peridot-green in colour, and showed two bands having nearly the same wavelength as the first two mentioned above, namely 684 and 660nm.

It must be admitted that such bands have little diagnostic

importance, but it is necessary for the practical gemmologist to realize that the appearance of chromium bands, unless these be sharply defined, does not necessarily denote that the material examined is a natural stone.

Didymium glass

Pink glasses owing their colour to didymium – the combined rare earth elements, praseodymium and neodymium – are sometimes used to imitate gems. These have a striking absorption spectrum of narrow lines, particularly a group in the yellow with its strongest members at 584 and 572nm. A drawing of this spectrum was printed in Chapter 12, and a general discussion of didymium spectra in gems can be found in Chapter 32.

Two other coloured glasses have shown didymium absorption strongly, the first a deep brick red which may have had an additional colorant with the didymium; and the second a dark brown in which the absorption forms broad black bands that masked the individual lines.

The only gemstone showing didymium lines of comparable strength to the pink glass is apatite; and the colour and appearance of the glass should be sufficient to prevent confusion here. Actually, the appearance of the spectrum is different in the two cases. In apatite, the group of fine lines in the yellow form an almost continuous 'block', while in didymium glass the 584 and 572 lines stand out prominently, with a marked gap in between them. The type of glass in which didymium is found seems to be the normal crown glass with refractive index near 1.52: other samples containing much more than usual of the rare earth gave R.I. of 1.543; 1.57 (dark red); and 1.63 (brown), the latter obviously being 'flint' glass. These three all gave strong spectra, with the main group in the yellow forming a solid block from about 600 to 566nm and three or more narrow bands or blocks of bands distributed through the spectrum to the violet. All were spectacular in their performance.

An interesting use was found for didymium glass in the 1914-18 War. Dot-dash messages were transmitted from an apparently steady light-source (thus not attracting enemy attention) by passing a didymium glass filter across the beam. The signals were received by an observer using a field-glass with prism attachment through which the intermittent appearance of a dark line in the yellow part of the spectrum for varying lengths of time could be seen as a morse message.

Selenium glasses

While red glasses incorporating colloidal gold are well-known in the history of art and decoration they are difficult to prepare and hardly a commercial proposition for imitation gems. Red glasses produced by copper under reducing conditions are so deep in tint that they are used by 'flashing' a thin film on to the surface of a colourless glass base. The most successful and common of the pink and red glasses used as gem substitutes probably owe their colour to selenium used in conjunction with cadmium sulphide. A broad band in the green is the only obvious feature of such glasses, and since this is also a usual feature in various red gemstones, it has very little value in practice. This is the more so because of a wide variation in the position of this central band. Measurements of 532, 537, 540 and 560nm, have been made in different specimens. It seems unlikely that all of these can owe their colour to the same cause.

Orange glasses often show a fierce cut-off below 590nm, beyond which wavelength they are virtually opaque. Fire opal, which resembles these orange glasses in colour, usually has a rather less definite cut-off and may transmit light down to 575 or even to 555nm, according to its depth of tint. The very low refractive index of the fire opal is the best check here: values in the 1.44 region are not found in any imitation gem.

Uranium glass

Fluorescent 'canary' yellow or green uranium glass was popular in Victorian times for the fabrication of small ornamental articles. Occasionally such glasses are found cut as imitation gems. The refractive index of specimens we have encountered is rather low (1.48). These are most easily distinguished by their banded fluorescence under ultra-violet light, but, whereas pure uranium salts show a banded *absorption* in the green also, such bands were not detected in the glasses we have seen. There are, however, vague bands in the blue and violet, which gave measurements of 495, 460 and 430nm, the last-named band acting as a cut-off to the violet. Very similar bands were found in a bead of fused microcosmic salt into which a fragment of uranyl nitrate had been introduced.

It will be realized from the above somewhat sketchy account of the absorption bands that we have noted in coloured glasses used as gems that only in the cobalt blue glass can the spectroscope be said to provide really positive evidence of the nature of the material. More often, the signs are merely negative. For instance, when a 'stone' of peridot colour shows no trace of the peridot bands in the blue, there is at least a strong probability that one is dealing with a paste, since no natural stone other than peridot displays quite this shade of green.

38   Fluorescence Spectra

Our survey of the spectroscope and of the ways in which it can serve the gemmologist is now virtually concluded; but before summarizing and indexing our scattered data on absorption spectra we propose to devote one last chapter to results we have obtained in examining with the spectroscope the fluorescent light emitted by several of the gem minerals when exposed to ultra-violet light, X-rays, or strong visible light. In most cases the fluorescent glow is too weak to yield results; in others the spectroscope reveals only a dim stretch of continuous spectrum. But in some instances the fluorescence spectrum shows discontinuities which are characteristic of the substance examined, and in a few minerals one is rewarded by the sight of definite bright lines or bands which are not only beautiful in themselves but have considerable diagnostic value.

The study of the luminescence spectra of minerals may be said to have preceded that of their absorption spectra, since as early as 1859 Edmond Becquerel gave a description, illustrated by excellent drawings, of the spectra emitted by minerals viewed in his ingenious 'phosphoroscope'.[1] Becquerel's drawings show the position of the bright bands he observed against a framework of the main Fraunhofer lines, so that their wavelength can be roughly assessed. No comprehensive survey of the kind seems to have been attempted in more modern times, though intensive detailed work has been carried out – by Deutschbein, for instance, on the chromium-activated phosphors,[2] and Raman and his school on the luminescence of diamond.[3]

In the notes which follow we can make no claim to completeness, since this aspect of spectroscopy has not been rigorously studied in our laboratory. But some of the data given may help to make the study of luminescence in gemstones of greater interest and value.

Activating elements

We have seen how, in coloured minerals the elements giving rise to colour are usually present only in very small amounts in an otherwise colourless host crystal. So it is with luminescent minerals. Pure minerals or chemical compounds seldom show any fluorescence; luminescent effects are most commonly due to mere traces of 'activators' in the otherwise inert substance of the 'phosphor'.

A large number of 'activating' elements are known; but only a few of these can give rise to a 'line' or banded fluorescence – the condition being that not the outer valency electrons, but those from an inner shell must be involved. Among such elements, chief in importance to the gemmologist is chromium, and the more important chromium fluorescence spectra have inevitably been described already in this book, since they appear simultaneously with absorption lines in such cases as ruby and spinel and can hardly be treated separately from these. Traces of several of the 'rare earth' elements can also give rise to a luminescence spectrum showing bright lines, and such spectra have played an important part in controlling the difficult chemical separation of these closely related elements. In the fluorescence spectrum of one or two gem minerals rare-earth lines can be detected. The green fluorescence emitted by salts and substances containing uranium reveals through the spectroscope a series of from five to seven bright bands which are entirely characteristic of this element. In addition to these cases we have in diamond the unique phenomenon of a banded fluorescence due to included nitrogen atoms which distort the diamond lattice itself.

Chromium fluorescence

In minerals containing chromium as colouring agent a red fluorescence can usually be stimulated by ultra-violet or even by ordinary visible light. For this to happen, however, two conditions must be satisfied: 1. The chromium must be properly 'built-in' to the lattice; this is usually the case when one is dealing with minerals containing alumina as an essential constituent, since in these chromic oxide can replace some of the alumina in the lattice. 2. Iron must not be present in appreciable amount, as this element has a marked quenching effect on luminescence.

For observing the red fluorescence due to chromium the Stokes 'crossed filter' technique, which has been described in Chapter 11, is incomparably the most powerful and appropriate.[4] Light passing through copper sulphate solution, or its equivalent, on to a gemstone is examined through a red filter. A chromium-activated mineral is stimulated by the blue light to fluoresce in the red, which is then seen against a dead black background through the red filter.

In ruby and red spinel the effect is brilliant and strong, and the two are indistinguishable to the unaided eye. But the spectroscope reveals the very different patterns of bright lines involved in this red luminescence. The 'organ-pipe' structure of the red spinel fluorescence is a most sensitive diagnostic feature, and is particularly valuable in a material which shows few absorption lines. It is quite distinct from the ruby fluorescence in which the predominant light comes from the main doublet only. Pink topaz and alexandrite show a feeble glow, in which the main doublet is seen as an emission line. Some emeralds glow strongly, and Chatham, Lechtleitner and Linde synthetic emeralds do so more strongly than most natural stones. It is difficult to see the emission doublet in this spectrum, which photography has shown to consist mainly of broader bands in the very deep red. Kyanite and hiddenite are among other chromium-activated minerals which show a red glow under crossed filters. The presence of iron prevents any fluorescence being shown by pyrope, demantoid or jadeite, although chromium is largely responsible for the attractive colour of these stones.

Rare earth spectra

Apart from the chromium-activated minerals the most exciting line fluorescence spectrum found amongst the gemstones is to be seen in some yellow, brown and fired colourless zircons. These show quite a bright yellow fluorescence under ultra-violet light, and the spectroscope analyses this into two groups of three narrow bands – one group in the yellow at 580, 574 and 570nm; the other in the blue at 485, 480 and 473nm. Not all such stones fluoresce and, in those that do, the emission lines are not easy to see and no great accuracy is claimed for these measurements, which are difficult to take with so dim a light-source. A drawing of these bright lines is reproduced in Fig. 38.1. This beautiful spectrum is certainly ascribable to one or more of the rare earth elements. Unfortunately, in the excitement of examining this bright line effect, white zircons may stay too long under the rays, and become discoloured, assuming a brownish tinge. One can usually restore the stones to their colourless state by heating to dull redness. This gives an opportunity for observing the thermophosphorescence, in which a green band makes its appearance between the yellow and blue groups. As heating is continued the yellow group is the first to disappear, leaving the green and blue lines to persist until the stone has become completely decolorized, when phosphorescence ceases.

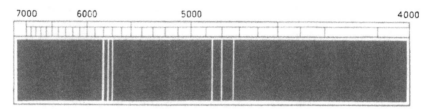

Fig. 38.1 Fluorescence spectrum of WHITE ZIRCON

A somewhat similar line spectrum can be seen in the violet-blue fluorescence of certain fluorspars; more particularly when stimulated by X-rays. The fluorspar is then strongly phosphorescent, and also thermophosphorescent. Under ultra-violet light (365nm) the line spectrum in most specimens we have examined was difficult to detect. Instead, a continuous spectrum extending from 525 to a sharp cut-off at 410nm is seen. No visible phosphorescence can be seen after exposure to this radiation, but there is phosphorescence in the ultra-violet, as can be proved by leaving the specimen on a photographic film for a few hours. The film, when developed, shows quite a strong image of the face of the specimen with which it was in contact (Fig. 38.2).

Yellow apatites from Mexico or Spain also show a line spectrum in the dull bluish glow they emit under ultra-violet light. These lines are probably not due to the didymium elements which give rise to the absorption spectrum, but to other rare earths of this group.

Other banded fluorescence

The banded uranium fluorescence mentioned earlier is easily seen in the yellow-green 'canary' glass once popular as a material for small ornamental objects; and this glass has sometimes been used in faceted form as an imitation gem. The brilliantly yellow-fluorescing scapolite from Canada, which is a favourite with those who stage displays of fluorescent minerals, also reveals this banded structure, but is now known to be caused by sulphur.

Fig. 38.2 Auto-photograph of FLUORSPAR crystal exhibiting ultra-violet phosphorescence

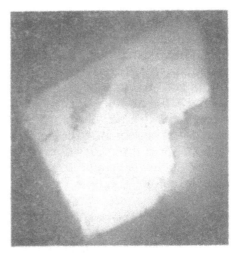

Another substance famed for its fluorescence is the lilac kunzite variety of spodumene, which gives an orange glow under ultra-violet rays. Under X-rays the effect is far stronger, and faint bright lines can then be seen on a background of continuous spectrum. There is then a persistent phosphorescence and (if irradiated for some time) a photo-coloration to green. The colour of the irradiated specimen is restored by heating, the process being accompanied by a strong thermophosphorescence.

Another favourite mineral for fluorescent displays is the aragonite from Girgenti, Sicily. Here the rosy pink luminescence conceals a green component which can, however, be seen in the phosphorescent afterglow which persists for some twelve seconds after removal from the ultra-violet beam. While the specimen is glowing under the rays, the spectroscope reveals a strong orange band of light extending from about 615 to 590nm. A fainter continuous spectrum begins at 545nm and extends into the blue, a dark region separating the two. When the rays are cut off the orange component vanishes at once, leaving only the green component as a short afterglow. Some calcites also display a red luminescence which shows a similar structure under the spectroscope, but here both components fade immediately the mineral is removed from the rays.

Diamond fluorescence

The effects just described are scientifically interesting; the luminescence of diamond, now to be considered, is even more interesting and also of more value in practical testing and identification. Diamond has been popularly credited with showing many different luminescent colours, but in fact one can reduce these to three broad categories only: blue-fluorescent, yellow-fluorescent and green-fluorescent stones.

The vast majority of diamonds, particularly those of fine gem quality, are blue-fluorescent under ultra-violet rays, varying from a bright whitish sky-blue through dimmer blues and blue-violet tints, to those which by comparison seem quite inert under the rays. Of some thousands of crystals of top quality examined only two or three were found to show the second most common luminescent colour – yellow. In industrial stones the proportion of yellows is much higher, while only in stones of the 'brown' series is a green glow seen with any frequency. The blue fluorescence reveals under the spectroscope a series of five or six bright bands, the strongest of which is in the same position as the main absorption band, at 415nm, and the others evenly spaced, or nearly so at 422, 433 (wavelengths only approximate) and on into the blue. Completely dark regions of non-fluorescence are seen to separate the deep violet bands. There is a continuous spectrum extending faintly all the way up to the red, though this is very feeble. The yellow-fluorescers show a continuous spectrum between about 625 and 490nm and then very faintly extending into the blue, while the green-fluorescent stones show a continuous region extending from 590 to 503nm and another dimmer region from 460 into the violet, while between the two is a marked dark region crossed by a few bright lines, difficult to observe, which belong to the 504nm absorption series, seen in reverse. By means of photography, using long exposures and special conditions, Professor Raman's team of physicists at Bangalore have revealed a far more detailed series of fluorescent lines and bands than those sketched above. But the description is of the effects seen under ordinary conditions with the prism spectroscope, and is at least sufficiently distinctive to enable a blue-fluorescent stone, which might be one of several things, to be identified with certainty as a diamond.

All blue-fluorescent diamonds show a rather persistent yellow afterglow, the strength of which is strictly proportional to the intensity of the preceding blue fluorescence. The effect can best be detected by holding the stone as close as possible to the rays for a short time in a darkened room, allowing the eyes to become dark-adapted and therefore sensitive to the faintest glimmer of light. The stone is then taken away from the rays very quickly and viewed at once within cupped hands, with the eye applied to the only orifice in the 'cup'. In this miniature 'dark-room' even a feeble glow can be detected, and this simple procedure forms a useful test for diamond, whether mounted or unmounted. The fact that the eye is too close to focus on the stone does not matter; it is the glow that is important.

A great deal of research is being done in various centres, on the cause and nature of luminescence in diamond. In the above brief account we have only concerned ourselves with the visual effects seen through a small prism spectroscope.

The light emitted by fluorescing minerals is at best rather feeble, and this renders details of the spectrum of this light difficult to observe and still more difficult to measure. A quartz or glass condensing lens is almost essential to focus the rays on the specimen and thus increase the brightness of the emission. In this chapter the term 'ultra-violet light' refers to the 365nm mercury radiation filtered through a Wood's glass filter. Effects seen with a short-wave mercury lamp are as a rule too feeble to lend themselves to analysis with the spectroscope.

Fluorescence and the laser

It will be seen that, apart from the fluorescence of ruby and other chromium gems which have been discussed at some length, by far the larger and more valuable section of this book has concerned itself almost exclusively with absorption, and the original papers here dealt with fluorescence rather as an afterthought, for the sake of completeness.

Anderson and Payne, in 1955, could not possibly have foreseen that, in five years time, the phenomenon of fluorescence – useful enough in gemmology, but otherwise little more than a pleasing visual curiosity – would leap to the forefront of scientific discovery with the construction of the first laser, which provided science with a new and dynamic kind of light or heat energy in an unimagined controllable form, limited only by available fluorescent wavelengths.

Lasers have been mentioned before in this edited and updated version of the papers, and it seems to me that a short description of them makes a fitting finale to the book.

The first 'laser' (Light Amplification by Stimulated Emission of Radiation) was made from a cylindrical rod of synthetic ruby, with optically polished plane and parallel surfaces at either end, one silvered to provide a full mirror and the other only partially reflecting. An intense light source was supplied by wrapping the equivalent of an electronic photographic flash unit around the rod. When this was discharged the chromium atoms in the ruby caused it to fluoresce strongly. Much of the fluorescent light was lost again through the sides of the rod, but some photons passed parallel to its length, reflecting an infinite number of times from the mirrored ends so that non-parallel beams were lost off the mirror edges, the remainder striking the chromium atoms repeatedly, increasing the fluorescent effect with each reflected passage.

The whole process takes perhaps a thousandth of a second or less, but in a rod of a few centimetres in length and with a light speed in ruby of about 170,000 kilometres per second it can be seen that the number of reflections back and forth must run into millions even in that short space of time, until the comparatively small percentage (10 per cent has been quoted), which is allowed to leak through the partially silvered end, builds up to an immensely powerful narrow, parallel beam of red light.

This beam has two features which are unique in optical science; first the entire beam is strictly in phase (as against normal light in which the phase is totally random) and secondly the light is practically of a single wavelength, monochromatic to a degree which would be impossible by other means.

Another remarkable feature is that such a beam has little tendency to diverge or spread and will remain parallel over immense distances. A laser beam directed at a static reflector left on the Moon was shown to have spread to only about 400 metres diameter after a round trip of about 750,000 kilometres. (This measured the distance of the Moon accurately to within 15cm.)

Since that first simple ruby laser other suitable materials have proliferated, and we now have a great variety of solid (pulsed) lasers, gas lasers and liquid lasers, and the invention which at first was looking for a use, is now a vital part of many aspects of our daily lives, of industry and of science. Among the multitude of applications today in one or other of its many forms are: high-speed tills, plastic card identification, ticket reading and other processing; information storage, retrieval, holography, printing, CD disc playing, micro-machining, automatic cutting and welding, laser surgery (self-cauterizing), weapon aiming, communication systems in conjunction with fibre optics, measuring and aligning, pollution monitoring, and even lasers to activate other lasers. Their uses for the future are beyond imagining.

The laser does not impinge greatly upon gemmology, but to ignore it here would be to pass up on one of the greatest of optical discoveries. The search for laser crystals has been responsible for the introduction of many new synthetic gems in recent years, either to the collectors' market, or in a few instances to that of the normal gem trade. The laser also contributes to the long story of diamond, which until quite recently could be cleaved only parallel to an octahedral face, or sawn with some difficulty parallel to the cube face. Today a heat laser can 'saw' even heavily twinned diamond in any direction. It may also be used to vaporize or burn holes of hair-like diameter from the surface of the diamond to an offending black inclusion which is then said either to be leached out by acid, or combusted by the heat of the laser probe. Or air is allowed to enter between the diamond host and the dark inclusion, which then reflects as a more acceptable white one.

I feel that the gem trade can take some pride in the fact that synthetic ruby, one of their products, was in at the start of this fantastic industry.

With this slight digression from spectroscopy we bring this account to a close. A summary of the most important spectra grouped by stone colour will now be given.

39 Summary of Absorption Spectra

A great many chapters have been needed to deal at all adequately with the applications of the spectroscope to gemmology: a fact which would have seemed incredible in the first decades of this century. The great bulk of information has been concerned with our own investigation of the absorption spectra of gem materials, for this is the practical application of the instrument to this specialized subject. We now provide a summary of the important features of each spectrum which should be quite adequate for diagnostic purposes.

Practical gemmologists use colour extensively as a guide, and each colour variety of most species usually has a different absorption spectrum from other colour varieties of the same species, so the stones described will be classified on a colour basis, proceeding in spectrum order, dealing with red, orange, yellow, green and blue stones in turn. Where it is known, the colouring element is indicated by its chemical symbol and valency in parenthesis.

In order to give some indication of the relative reliability and practical importance of each spectrum a 'star' system has been adopted. The significance of these symbols is as follows:

A 'three-star' spectrum is one which is completely diagnostic and invariably visible in the stone concerned.

A 'two-star' spectrum is one which may not always be fully visible but which is diagnostic when well-developed and clearly seen.

A 'one-star' spectrum is one which is not very reliable or easy to interpret, but which is of occasional value as a confirmatory test.

Colourless stones

Zircon *** (U''). A narrow line in the red at 653.5nm is invariably visible, accompanied by a fainter companion line at 659nm. By

248

transmitted light these lines may be faint and hard to see, but by internally reflected light they should be visible even to the beginner. Others of the stronger bands of the zircon spectrum (given under 'yellow stones') in the yellow and green are often discernible.

Diamond ** (N'''''). A narrow band in the deep violet at 415.5nm can usually be detected. Cape stones show this band strongly as well as more diffuse bands in the blue and violet, of which the strongest is at 478nm.

The important spectra associated with diamonds coloured by irradiation have been dealt with in Chapter 33, and the reader is referred to that chapter when dealing with coloured diamonds rather than attempt to summarize this complex subject here.

Rutile (synthetic) **. Synthetic rutile is never quite free from a yellowish tint, but perhaps may be included here, since it has been used to imitate diamond. The spectrum ends sharply at 425nm where an intense absorption band begins, extending into the ultraviolet.

Red and pink stones

Ruby *** (Cr'''). A close doublet in the deep red at 694.2 and 692.8nm, usually seen as a bright fluorescence line, is very distinctive for corundum containing traces of chromium. In ruby there are weaker lines at 668 and 659.5nm, a broad band in the green centred near 550nm, and three strong lines in the blue at 476.5, 475 and 468.5nm. Only the important lines are mentioned here in an unusually rich chromium spectrum. The fluorescence line and the three lines in the blue (two very close together, the third rather apart from these) are completely diagnostic for ruby. Synthetic ruby has the same spectrum, somewhat enhanced by the rather greater chromium content, and absence of iron.

Spinel ** (Cr'''). Narrow absorption bands in the red are seldom seen. More reliable, and very characteristic, is a group of five or more narrow red fluorescence lines (the strongest of which is at 685.5nm), best seen by scattered light. There is a broad absorption band centred at 540nm in the green (compare pyrope), but, unlike ruby, there are no easily observed lines in the blue.

Pyrope ** (Cr''', Fe''). Narrow chromium lines may be seen in the red, but the diagnostic features are a broad band in the yellow-green, centred at 575nm (compare spinel) and the strongest of the almandine bands, at 505nm in the blue-green, weakly developed. There is no fluorescence.

Almandine *** (Fe''). A distinctive spectrum of three strong broad bands in the yellow at 576, green at 527, and blue-green at 505nm with weaker bands in the orange, 617 and blue, 462.

Tourmaline ** (Mn)? A diffuse band in the green is centred near 525 with a distinctive fine line at 537nm near the long-wave end of this. Two narrow bands in the blue at 458 and 450nm are also distinctive. There is no fluorescence.

Topaz * (Cr'''). Fired pink topaz shows a faint line at 682nm in the red, either as an absorption or as a fluorescence line.

Zircon ** (U''). Red zircon may show a strong spectrum (as under 'yellow') but some red zircons show virtually no spectrum.

Orange stones

Spessartite *** (Mn''). Often shows faint almandine bands, but in addition shows distinctive manganese bands at 485 and 461nm in the blue, with a very strong band at 430nm in the violet.

Of other orange stones, fire opal transmits only orange and some red light, hessonite has no distinctive spectrum of its own, though some specimens may reveal traces of the almandine or the spessartite spectrum.

Scheelite *** (Dm). In its golden orange colour, although a collectors' gem and rarely cut, it has a strong didymium spectrum rivalling that of the classic example of yellow apatite.

Corundum **. The strange and very rare orange-red sapphire known as padparadschah and found in Sri Lanka, has a chromium fluorescent line at 635nm, as in ruby. Synthetic 'padparadschahs' show this even more strongly.

Yellow stones

Yellow zircon *** (U''). Usually shows all the strongest lines of the uranous spectrum of zircon, including 691, 662.5, 659, 653.5 in the red, 589.5 in the yellow (note that this coincides with the position of the sodium doublet emitted by a sodium flame or discharge lamp) 562.5, 537.5 and 515 in the green, 484 and 432.5nm in the blue and violet.

Yellow sapphire ** (Fe''). Yellow sapphires from Australia, Thailand or Montana show a group of three bands in the blue, the strongest at 450, the others at 460 and 471nm – each being broad enough to be almost in contact with the others. Sri Lankan yellow sapphires contain very little iron, and show only the 450nm faintly, or no bands at all. Synthetic yellow sapphires never show any of these absorption bands, neither do they show the yellow fluorescence under ultra-violet light seen in Sri Lankan yellows.

Chrysoberyl *** (Fe''). A single broad band is seen in the blue-violet centred at 445nm. Apart from its position this band is distinguishable from those in yellow sapphire in being a single block and not a complex.

Yellow orthoclase ** (Fe''). Rather diffuse bands are seen at 448 and 420nm, in the violet.

Yellow spodumene ** (Fe''). A narrow band is seen in the violet at 438 with a weaker companion at 432.5nm. Jadeite has a very similar spectrum, but the gemstones are so different in appearance that no confusion should be possible.

Yellow apatite, danburite, and sphene (Didymium). These three uncommon yellow stones can be grouped together, since they all show the typical lines in the yellow indicating didymium, though in very different strength. In apatite there is a strong group of fine lines in the yellow region, of which the strongest is near 584nm. There is a weaker group in the green near 527nm. Yellow apatite can perhaps claim to have a 'two star' spectrum, since the very faint lines in the yellow which can be detected in most danburites and sphenes by internally reflected light are unlikely to be confused with the far more robust bands in apatite.

Sinhalite ** (Fe''). Although usually associated with peridot, this is rarely a green stone, and may occur in various shades of yellow. The distinctive feature is a band at 463nm between the 475 and 452 bands which are normal to peridot. Bands at 493 and 562nm complete the resemblance to the latter spectrum, but in this case their colours are quite different.

Green stones

Emerald *** (Cr'''). Emerald shows a typical chromium absorption spectrum, with a strong doublet in the red at 683.5 and 680.5nm. Weaker and more diffuse lines occur at 662 and 646nm, with a curious patch of high transparency beside each. There is also a line at 637nm, which in the ordinary ray is almost as intense as the doublet. A broad absorption band in the yellow and yellow-green is rather weak. In fine specimens a line at 477 is seen in the ordinary ray.

Alexandrite *** (Cr'''). A chromium spectrum more clear-cut than in emerald – in fact these alexandrite lines are the sharpest in the chromium vocabulary. The strong doublet is at 683.5 and 678.5nm, and is only seen as a fluorescence line in very exceptional circumstances. Weaker lines are seen at 665, 655, 645nm. The broad absorption band is centred at 580nm in the yellow, and there are narrow lines in the blue at 473 and 468nm.

Demantoid *** (Cr''', Fe''). The finest demantoids owe their rich colour to chromium, and a doublet at 701 with diffuse lines at 640 and 621nm are then to be seen. But the most distinctive feature of demantoid or of yellow andradite garnet is an intensely powerful absorption band in the beginning of the violet, centred at 443nm. In deep-coloured specimens this may appear as a 'cut-off' to the end of the spectrum.

Green sapphire *** (Fe'''). An important spectrum, as it forms a most useful basis for distinguishing between synthetic and natural green, yellow and blue sapphires. A group of three strong bands in the blue almost coalesce, but the weakest band at 471nm is sufficiently detached to reveal that the band is a complex one. The strongest band, at 450nm, is seen in almost all natural blue sapphires (see also under yellow sapphire).

Peridot *** (Fe''). Three bands, rather diffuse, but with narrow 'cores', are evenly spaced in the blue, forming a distinctive pattern. The wavelengths are 493, 473 and 453nm. The strength and appearance of the bands varies in the three vibration-directions of light corresponding with the three principal refractive indices, being strongest in the 'beta' spectrum, where an additional band in the green at 529nm may be detected.

Sinhalite ** (Fe''). This is rarely more than greenish brown, but in that colour might be confused with the rare brown peridot, so it is included under this section. The distinctive feature is a band at 463nm, in the gap between bands at 475 and 452nm, which closely resemble those in peridot. A band at 493 and a weak band at 526nm complete the resemblance, apart from the 'extra' band mentioned above.

252

Chrysoberyl *** (Fe'''). A broad band at 445nm, as in yellow chrysoberyl, only stronger. Useful as a check for chrysoberyl cat's-eyes.

Tourmaline ** (Fe). There is complete absorption of the red down to approximately 640nm, and a narrow band can be usually seen at 497nm, where the green merges into the blue.

Aquamarine ** (Fe'''). The green type of aquamarine, and several other members of the beryl family show a narrow line at 537nm in the green in the spectrum of the extraordinary ray only. This is very faint, but distinctive where seen.

Enstatite *** (Fe''). A strong and narrow band at 506nm, between the green and the blue, is highly distinctive. Weak chromium lines may also be seen in fine-coloured specimens.

Chrome diopside ** (Fe''). Shows a spectrum similar to enstatite, but with a double line at 508 and 505nm in place of the single line of enstatite.

Zircon *** (U''). Green zircon always shows a distinctive spectrum, though this may vary considerably with the degree of metamictization. Where the internal breakdown is not complete, there are ten or eleven bands seen spaced throughout the spectrum (see yellow zircon for wavelengths) though these are noticeably 'woolly' compared with the corresponding lines in high zircon. Completely metamict zircons may show only a diffuse band at about 653nm in the red, though in certain types a narrow band in the green at 520nm can also be seen.

Epidote ** (Fe). Absorbs strongly from the end of the green onwards, but in sufficiently transparent stones a band at 475 can be seen, and a very intense band in the blue at 455nm.

Andalusite *. Brownish-green andalusites may show a band in the blue at 455nm. A bright green Brazilian type has a very beautiful rare-earth spectrum, with a knife sharp edge of shadow at 552.5nm in the green shading off to the yellow side with a narrow line at 549.5. Also a line nearer the blue at 517.5nm, and general absorption of the blue and violet.

Fluorspar *. Weak bands may be seen at 634 and 610nm in the orange, 582nm in the yellow and 446 and 427nm in the violet.

Kornerupine **. The spectrum of the 'beta' ray shows weak bands at 540, 503 and 463nm, with 446nm, rather strong, and finally 430nm, weak.

Jadeite ** (Cr''' Fe'''). The spectrum of fine green jadeite is a typical chromium one so far as the red is concerned, but more blurred than in translucent emerald, with which it might be confused. Where general absorption does not mask it, an intense and narrow band at 437nm, in the violet, is far more diagnostic and very useful, especially in the paler specimens of jadeite.

Nephrite can hardly be awarded a star. Chromium lines are sometimes vaguely seen, though the colour is mainly due to iron. Vague bands in actinolite at 510 and 495nm (both narrow) are sometimes detected in nephrite also.

Blue stones

Sapphire *** (Fe'''). This spectrum may not be invariably visible but is so important that we have awarded it three-star status. The three bands in the blue which belong to green sapphire are also seen strongly in blue sapphires from Australia. Natural sapphires from other localities show at least the strongest of these three, as a rather narrow line at 450nm. In some Sri Lankan stones the band is so faint as to be virtually absent. It is an 'ordinary ray' band, and may be enhanced by the use of a polaroid disc turned to the correct angle. Where clearly seen, the sapphire concerned is certainly natural, and this forms a valuable accessory test.

Spinel *** (Fe''). The bands in blue spinel are rather vague, but form a very distinctive pattern to the practised eye. There is a broad band in the blue centred at 459, with a narrow band nearer the green at 480nm. Fainter ones in the green (555) and yellow (592) and another in the orange (632) complete the pattern.

Aquamarine ** (Fe'''). A weak, vague band in the blue at 456 is followed by a stronger band in the violet at 427nm, the two together forming a convincing test for the gemstone in the large specimens which are frequently encountered. Maxixe aquamarine has an extraordinary spectrum with strong bands in the red at 697 and 657nm, and a weaker orange band at 616nm.

Iolite *. The spectrum of iolite varies greatly with direction, as might be expected in so pleochroic a gem. In the blue direction there are vague bands in the blue and violet at 492, 456 and 437nm. In the yellow ray a fairly narrow double band can be seen at 593 and 585nm, with a narrow line at 535nm.

Apatite *. In the yellowish ordinary ray of the blue apatite from

Burma a number of bands can be seen, the two strongest being a fairly narrow one at 511nm in the green and 490nm in the blue.

Turquoise ** (Cu). There are two similar absorption bands in the violet, but only one of these, a fairly narrow and strong one at 432nm in the violet, can commonly be seen. A vague band in the blue at 460nm is an additional feature and these two bands can usually be detected by reflected light in genuine turquoise, and are exceedingly useful in checking the authenticity of specimens of this mineral, for which so few satisfactory tests exist.

Tourmaline **. Blue tourmaline shows the 497nm band mentioned under 'green'.

Synthetic blue spinel *** (Co). This common synthetic shows a distinctive cobalt spectrum in strength varying with the depth of colour. There are three broad bands centred at 635, 580 and 540nm. These can be distinguished from the bands in cobalt glass by the fact that the central band is broader than the others, whereas in glass this band is narrower than the flanking bands.

Blue cobalt glass *** (Co). Very similar to above, but the bands are at 655, 590 and 535nm. They are thus more widely separated, and, as just stated, the central band is the narrowest of the group.

Brown stones

These do not fall into the spectral range of colours, but there are enough of them with identifiable absorption spectra to justify a separate category here.

Zircon *** (U''). While reddish-brown zircons may show either a very faint line at 653.5nm, or none at all, other shades of brown from Sri Lanka may show up to about seven lines, a spectrum very similar to that of yellow zircon from that country. Some Burmese stones of brown or brownish green will give well-developed spectra of thirty or more lines.

Sinhalite *** (Fe''). This absorption has been described in detail under green stones, but the species is more normally found in shades ranging from straw yellow to deep brown. This affects absorption only in degree. Any possible confusion with brownish peridot can be resolved by noting that the R.I. of sinhalite is decidedly negative biaxial in sign, while gem varieties of peridot are positive.

255

Enstatite *** (Fe''). The rich brown stones from India give a very intense version of the spectrum described for the green variety of this stone, the 506nm line often being supported by a much fainter sharp line on either side.

References and Notes

3 The Absorption Spectra of Solids

1 G.G. Stokes, *Phil. Trans.*, 1852, Vol. 143, pp. 463–562
2 This paper is reproduced complete in an article by A.E. Farn, *Journ. Gem.*, 1951, Vol. 3, pp. 142–4
3 H.C. Sorby, *Proc. Roy. Soc.*, 1869, Vol. 17, pp. 511–15
4 E. Becquerel, *Annales de Chimie et de Physique*, 1859, Vol. 57, p. 40

8 Absorption Spectra

Anderson and Payne, 'The Absorption Spectrum of Zircon', *The Gemmologist* (August 1939)

11 Absorption and Fluorescence Spectrum of Ruby

1 Becquerel, *Ann. Chim. Phys.*, 1859, pp. 40–124
2 Crookes, *Nature*, 1888/9, pp. 537–43
3 de Boisbaudron, *Compt. Pend.*, 1886, p. 1107
4 Miethe, *Verh. Deutsch. Phys. Gesell.*, 1907, p. 715
5 Keeley, *Proc. Acad. Nat. Sci. Philadelphia*, 1911, p. 106
6 Dubois and Elias, *Ann. der Phys.*, 1911, p. 620
7 Mendenhall and Wood, *Phil. Mag.*, 1915, p. 316
8 Deutschbein, *Ann. der Phys.*, 1932, p. 712
9 Anderson and Payne, *Gemmologist*, 1948, p. 246

12 Absorption and Fluorescence Spectrum of Red Spinel

1 Becquerel, *Ann. Chim. Phys.*, 1859, pp. 40–124
2 Wherry, *Amer. Min.*, 1929, p. 324
3 Moir, *Trans. Roy. Soc. S. Africa*, 1912, p. 271

4   Deutschbein, *Ann. der Phys.*, 1932, p. 730
5   Gübelin, *Gems and Gemmology*, Winter 1952–3, p. 239
6   Anderson, *Gemmologist*, 1953, p. 42

14   Absorption Spectrum of Emerald

1   Moir, *Trans. Roy. Soc. S. Africa*, 1912
2   Deutschbein, *Ann. der Phys.*, 1932
3   Anderson, *Gemmologist*, May, 1935
4   Anderson, *Gemmologist*, 1953, p. 41

15   The Absorption Spectrum of Alexandrite

1   Deutschbein, *Ann. der Phys.*, 1932
2   Wherry, *Amer. Min.*, 1929

17   Absorption Spectrum of Demantoid Garnet

1   B.W. Anderson, 'An unorthodox test for demantoid', *The Gemmologist*, December 1942

18   Other Chromium Spectra

1   Aurousseau and Merwin, *Amer. Min.*, 1928, p. 559

19   Absorption Spectrum of Almandine Garnet

1   A.H. Church, *Intellectual Observer*, 1866, Vol. 9, p. 291
2   A.E. Farn, *Journal of Gemmology*, 1951, Vol. 3, p. 142
3   Anderson and Payne, *The Gemmologist*, 1954, Vol. 23, p. 163

20   Absorption Spectrum of Blue Spinel

1   E.T. Wherry, *American Mineralogist*, 1929, p. 326
2   R. Keith Mitchell, *Journal of Gemmology*, 1977, p. 356–7
3   J.E. Shigley and C.M. Stockton, *Gems & Gemmology*, 1984, pp. 34–41

21   Absorption Spectra of Peridot and Sinhalite

1   G.F. Claringbull and M.H. Hey, 'Sinhalite, a new mineral', *Min. Mag.*, 1952, Vol. 29, p. 841; see also C.J. Payne, 'Sinhalite – a new mineral and gemstone', *The Gemmologist*, 1952, Vol. 21, p. 177

22 Absorption Spectra of Enstatite and Diopside

1 R. Keith Mitchell, A new variety of gem enstatite, *The Gemmologist*, Vol. 22, 1953, p. 145
2 R. Keith Mitchell, 'Further notes on hypersthene-enstatite', *The Gemmologist*, Vol. 23, 1954, pp. 195–6

24 Absorption Spectra of Green Tourmaline and Iolite

1 J.E.S. Bradley and O. Bradley, 'Observations on the colouring of pink and green zoned tourmaline', *Mineralogical Magazine*, 1953, Vol. 30, pp. 26–38

27 Absorption Spectra of Aquamarine and Orthoclase: Spodumene and Jadeite

1 Bayley, *Physical Review*, 1927, Vol. 29, p. 353

28 Absorption Spectra of Andradite and Epidote

1 H. Becquerel: *Compt. Rend*, 1889, Vol. 108, p. 282

30 Absorption Spectra Due to Cobalt and Vanadium

1 In the original paper Mr Anderson was very firmly convinced that no natural blue gemstone could show the cobalt bands. In the last few years some doubt has been expressed and it now seems probable that certain natural spinels of quite exceptional blue colour do, in fact, owe that shade to the presence of minute quantities of cobalt.

32 Absorption Spectra of Zircon

1 Sorby, H.C., *Proc. Roy. Soc.*, 1868–9, 17, pp. 511–15
2 Chudoba, K.F., *Gemmologist*, 1937, 7, pp. 193–6
3 Anderson, B.W. and Payne, C.J., *Gemmologist*, 1939, 9, pp. 1–5

33 The Rare Earth Elements

1 The original text used the spectrum of didymium chloride to illustrate the full development of Nd absorption of visible light.

While I was preparing Anderson's work for book form, Dr J.B. Nelson very kindly sent me a piece of YAG, specially doped with an optimum percentage of pure neodymium, which gives such a remarkably sharp and complete absorption spectrum that I have decided to use it in place of Anderson's original illustration.

No natural gem mineral is likely to show anything like the perfect development seen in the YAG and in most instances only the complex of lines in the yellow region will be seen and that usually as a blurred but instantly recognizable band. This will be seen best in yellow or green apatite or in orange scheelite.

The YAG filter is sold as a precision quality spectral wavelength standard covering visible and near UV regions for use in highly technical spectroscopy using spectrophotometers, and is available from McCrone Research Associates Ltd.

2   The original drawing of the apatite spectrum showed further lines at the beginning of the blue and an intense band at about 449nm. These have been copied in more than one publication, but here Anderson's text does not mention them and in checking a considerable number of stones in my own collection I have failed to see them. I have therefore illustrated only the two diagnostic groups that Anderson has described, and I have seen. With a narrow slit, careful focusing and ideal lighting the two groups may be made to resolve extra lines within their structure, but it is the groups themselves, with their sharp edges on the shortwave side, that suggest apatite regardless of their content of lines. In many apatites the lines tend to merge to give two sharply defined broad bands.

3   I have to add that while researching for this book, I examined several blue apatites from Mogok which gave S.G. of 3.18 and the same remarkable dichroism as Anderson's stones, but which had a moderate didymium spectrum, and no trace of the special one he has described. So we have to assume that not all Burmese blue apatite can be expected to yield this rare spectrum.

4   Walton, Sir James, 'An unusual emerald', *The Gemmologist*, June 1950, pp. 123–5

5   *Gems & Gemology*, Vol. 14, 1973, p. 148

6   *Gems & Gemology*, Vol. 18, 1982, p. 230

7   Crowningshield R., 'Dark-blue aquamarine', *Gems & Gemology*, XIV, p. 111

38 Fluorescence Spectra

1   Edmond Becquerel, *Annales de Chimie et de Physique*, 1859, p. 40
2   O. Deutschbein, *Annalen der Physik*, 1932, p. 712
3   Anna Mani, *Proc. Indian Acad. Sci.*, 1944, p. 231
    Sir C.V. Raman, ibid. p. 199
4   B.W. Anderson, *The Gemmologist*, 1953, p. 39

Index

Index

Uvarovite, 119

vanadium, 184
Verde Antique, 163
Verneuil, 133, 227, 231–3

Walter, B., 216
Walton, Sir James, 209
Webster, R., 43, 182
Wherry, E.T., 63–4, 109, 133, 144, 227
Williamsite, 163
Wollaston, 35

Wulfenite, 206
Yunnan jade, 112

Zantedeschi, Prof., 24
Zimbabwe (Rhodesia), 124
Zinc Blende, 223–4
Zircon, 32, 194–202, 241, 248, 250, 253, 255
    anomalous, 200
    Burmese, 197
    heat treated, 197–8
    low type, 196
    metamict, 196–7

Notes

DIAMONDS, 3RD EDITION
THE ANTOINETTE MATLINS BUYING GUIDE
How to Select, Buy, Care for & Enjoy Diamonds with Confidence and Knowledge
by Antoinette Matlins, PG, FGA

Practical, comprehensive, and easy to understand, this book includes price guides for old and new cuts and for fancy-color, treated and synthetic diamonds. **Explains in detail** how to read diamond grading reports and offers important advice for after buying a diamond. **The "unofficial bible" for all diamond buyers who want to get the most for their money.**

6 x 9, 240 pp, 12 full-color pages, with over 150 color and b/w photos and illus.; index

Quality Paperback Original, 978-0-943763-73-6 **$18.99**

COLORED GEMSTONES, 3RD EDITION
THE ANTOINETTE MATLINS BUYING GUIDE
How to Select, Buy, Care for & Enjoy Sapphires, Emeralds, Rubies and Other Colored Gems with Confidence and Knowledge
by Antoinette Matlins, PG, FGA

This practical, comprehensive, easy-to-understand guide **provides in depth** all the information you need to buy colored gems with confidence. Includes price guides for popular gems, opals and synthetic stones. Provides examples of gemstone grading reports and offers important advice for after buying a gemstone. **Shows anyone shopping for colored gemstones how to get the most for their money.**

6 x 9, 256 pp, 24 full-color pages, with over 200 color and b/w photos and illus.; index

Quality Paperback Original, 978-0-943763-72-9 **$18.99**

THE PEARL BOOK, 4TH EDITION
THE DEFINITIVE BUYING GUIDE
How to Select, Buy, Care for & Enjoy Pearls
by Antoinette Matlins, PG, FGA
COMPREHENSIVE • EASY TO READ • PRACTICAL

This comprehensive, authoritative guide tells readers everything they need to know about pearls to fully understand and appreciate them, and avoid any unexpected—and costly—disappointments, now and in future generations.

- A journey into the rich history and romance surrounding pearls.
- The five factors that determine pearl value & judging pearl quality.
- What to look for, what to look out for: How to spot fakes. Treatments.
- Differences between natural, cultured and imitation pearls, and ways to separate them.
- Comparisons of all types of pearls, in every size and color, from every pearl-producing country.

6 x 9, 248 pp, 16 full-color pages, with over 250 color and b/w photos and illus.; index

Quality Paperback, 978-0-943763-54-5 **$19.99**

OR CREDIT CARD ORDERS CALL 800-962-4544 (8:30AM–5:30PM EST Monday–Friday)
Available from your bookstore or directly from the publisher. **TRY YOUR BOOKSTORE FIRST.**

The "Unofficial Bible" for the Gem & Jewelry Buyer

AWARD WINNER

Book Club Selection

JEWELRY & GEMS
THE BUYING GUIDE, 7TH EDITION

Updated Retail Price Guides

How to Buy Diamonds, Pearls, Colored Gemstones, Gold & Jewelry with Confidence and Knowledge

by Antoinette Matlins, PG, FGA, *and* A. C. Bonanno, FGA, ASA, MG

—over 400,000 copies in print—

Learn the tricks of the trade from *insiders:* How to buy diamonds, pearls, precious and other popular colored gems with confidence and knowledge. More than just a buying guide . . . discover what's available and what choices you have, what determines quality as well as cost, what questions to ask before you buy and what to get in writing. Easy to read and understand. Excellent for staff training.

6 x 9, 352 pp, 16 full-color pages, with over 200 color and b/w photos and illus.; index

Quality Paperback, 978-0-943763-71-2 **$19.99**

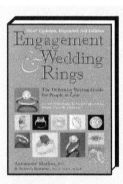

ENGAGEMENT & WEDDING RINGS, 3RD EDITION

by Antoinette Matlins, PG, FGA, *and* A. C. Bonanno, FGA, ASA, MGA

COMPREHENSIVE • EASY TO READ • PRACTICAL

Tells **everything you need to know to design, select, buy and enjoy that "perfect" ring** and to truly experience the wonder and excitement that should be part of it.

Updated, expanded, filled with valuable information.

Engagement & Wedding Rings, 3rd Ed., will help you make the *right* choice. You will discover romantic traditions behind engagement and wedding rings, how to select the right style and design for *you,* tricks to get what you want on a budget, ways to add new life to an "heirloom," what to do to protect yourself against fraud, and much more.

6 x 9, 320 pp, 16 full-color pages, with over 400 color and b/w photos and illus.; index

Quality Paperback Original, 978-0-943763-41-5 **$18.95**

JEWELRY & GEMS AT AUCTION

The Definitive Guide to Buying & Selling at the Auction House & on Internet Auction Sites

by Antoinette Matlins, PG, FGA

with contributions by Jill Newman

As buying and selling at auctions—both traditional auction houses and "virtual" Internet auctions—moves into the mainstream, **consumers need to know how to "play the game."** There are treasures to be had and money to be saved and made, but buying and selling at auction offers unique risks as well as unique opportunities. This book makes available—for the first time—detailed information on how to buy and sell jewelry and gems at auction without making costly mistakes.

6 x 9, 352 pp, 16 full-color pages, with over 150 color and b/w photos and illus.; index

Quality Paperback Original, 978-0-943763-29-3 **$19.95**

Now You Can Have the "Professional's Advantage!" With Your OWN Jeweler's Loupe—

The Essential "TOOL OF THE TRADE!"

Personally selected by the authors, this valuable jeweler's aid is *now available to the consumer* from GemStone Press. And GemStone Press includes, FREE, a copy of "The Professional's Advantage: How to Use the Loupe and What to Look For," a $5.00 value, written with the jewelry buyer in mind. You can now *have more fun while shopping and make your choice with greater confidence*. This is not just a magnifying glass. It is specially made to be used to examine jewelry. It will help you to—

- *Enjoy* the inner beauty of the gem as well as the outer beauty.
- *Protect yourself*—see scratches, chips, or cracks that reduce the value of a stone or make it vulnerable to greater damage.
- *Prevent loss*—spot weak prongs that may break and cause the stone to fall from the setting.
- *Avoid bad cutting*, poor proportioning and poor symmetry.
- *Identify the telltale signs* of glass or imitation.
- *. . . And much more, as revealed in "The Professional's Advantage!"*

You'll love it. You'll enjoy looking at gems and jewelry up close—it makes this special experience even more exciting. And sometimes, as one of our readers recently wrote:

"Just having the loupe and looking like I knew how to use it changed the way I was treated."

CALL NOW AND WE'LL RUSH THE LOUPE TO YOU.
FOR TOLL-FREE CREDIT CARD ORDERS:
800-962-4544

Item	Quantity	Price Each	TOTAL
Standard 10X Triplet Loupe	_____	$29.00	= $_____
Bausch & Lomb 10X Triplet Loupe	_____	$44.00	= $_____
"The Professional's Advantage" Booklet	1 per Loupe	$ 5.00	= ___Free__
Insurance/Packing/Shipping in the U.S.*	1st Loupe	$ 7.95	= $___7.95__
*Outside U.S.: Specify shipping method (insured) and provide a credit card number for payment.	Each add'l	$ 3.00	= $_____

TOTAL: $_____

Check enclosed for $_____ (Payable to: GEMSTONE PRESS)
Charge my credit card: ❑ Visa ❑ MasterCard
Name on Card _____
Cardholder Address: Street _____
City/State/Zip _____ E-mail _____
Credit Card # _____ Exp. Date _____
Signature _____ CID# _____
Please send to: ❑ Same as Above ❑ Address Below   Phone (____)_____
Name _____
Street _____
City/State/Zip _____ Phone (____)_____

TOTAL SATISFACTION GUARANTEE

If for any reason you're not completely delighted with your purchase, return it in resellable condition within 30 days for a full refund.

Phone, mail, fax, or e-mail orders to:
GEMSTONE PRESS, Sunset Farm Offices, Rte. 4, P.O. Box 237, Woodstock, VT 05091
Tel: (802) 457-4000 • *Fax:* (802) 457-4004 • *Credit Card Orders:* (800) 962-4544
sales@gemstonepress.com • www.gemstonepress.com
Generous Discounts on Quantity Orders

Prices subject to change

ASK ABOUT OTHER GEM ID EQUIPMENT — REFRACTOMETERS • DICHROSCOPES • MICROSCOPES • AND MORE

OR CREDIT CARD ORDERS CALL 800-962-4544 (8:30AM–5:30PM EST Monday–Friday)

Do You Really Know What You're Buying?
Is It Fake or Is It Real?

> The companion book to
> Jewelry & Gems:
> The Buying Guide

If You Aren't Sure, Order Now—

AWARD WINNER

GEM IDENTIFICATION MADE EASY

Antiques Roadshow Book Club Selection

UPDATED, REVISED, EXPANDED EDITION
THE ONLY BOOK OF ITS KIND
GEM IDENTIFICATION MADE EASY, 5TH EDITION
A Hands-On Guide to More Confident Buying & Selling
by Antoinette Matlins, PG, FGA, *and* A. C. Bonanno, FGA, ASA, MGA

The only book that explains in non-technical terms how to use pocket, portable and laboratory instruments to identify diamonds and colored gems and to separate them from imitations and "look-alikes."

The book's approach is direct and practical, and its style is **easy to understand.** In fact, with this easy-to-use guide, *anyone* can begin to master gem identification.

Using a simple, step-by-step system, the authors explain the proper use of essential but uncomplicated instruments that will do the identification tasks, what to look for gemstone-by-gemstone, and how to set up a basic lab at modest cost. **Three of the instruments are inexpensive, portable pocket instruments that, when used together, can identify almost 85% of all precious and popular stones.**

NEW Gems
NEW Treatments
NEW Synthetics
NEW Instruments
NEW Techniques

Including Complete and Easy Instructions:
◆ **Setting Up a Basic Lab**
◆ **Description of Each Instrument —**
What It Will Show & How to Use It
SSEF Diamond-Type Spotter • Electronic Diamond Dual Tester • Dark-field Loupe •
Synthetic Emerald Filters • Immersion Cell • Synthetic Diamond Detector
Loupe • Chelsea Filter • Refractometer • Ultraviolet Lamp •
Microscope • Spectroscope • Polariscope • Dichroscope
◆ **Antique and Estate Jewelry** — *The True Test for the Gem Detective*
Dyeing • Composite Stones • Foil Backing • Substitutions
◆ **Appendices:** Charts and Tables of Gemstone Properties, Schools, Laboratories,
Associations, Publications and Recommended Reading

As entertaining as it is informative. Essential for gem lovers, jewelers, antique dealers, collectors, investors and hobbyists. **"THE BOOK YOU CAN'T DO WITHOUT."** —*Rapaport Diamond Report*

6 x 9, 400 pp, with over 150 photos and illus., 80 in full color; index
Hardcover, 978-0-943763-90-3 **$38.99**

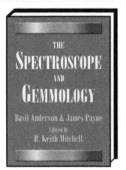

THE SPECTROSCOPE AND GEMMOLOGY

Basil Anderson & James Payne

Edited by R. Keith Mitchell

THE SPECTROSCOPE AND GEMMOLOGY
Ed. by R. Keith Mitchell, FGA
"Well written and illustrated. An invaluable work for anyone involved in gemstone identification."
—*Choice, Science & Technology*

The first book devoted exclusively to the spectroscope and its use in gemstone identification, this comprehensive reference includes the history and development of spectroscopy; discussion of the nature of absorption spectra and the absorption spectra of solids and gem minerals; the many uses of the spectroscope and the spectrophotometer; light sources and the causes of color; absorption in synthetic gemstones; and more. Indispensable for professional and amateur gemologists and students of gemology.

6 x 9, 280 pp, with over 75 b/w illus.; index
Quality Paperback, 978-0-943763-52-1 **$49.95**

FOR CREDIT CARD ORDERS CALL 800-962-4544 (8:30AM–5:30PM EST Monday–Friday
Available from your bookstore or directly from the publisher. TRY YOUR BOOKSTORE FIRST.

CAMEOS OLD & NEW, 4TH EDITION
by Anna M. Miller, GG, RMV; *revised, updated and expanded by*
Diana Jarrett, GG, RMV

Newly updated and expanded, *Cameos Old & New*, 4th Ed., is a **concise, easy-to-understand guide** enabling anyone—from beginner to antique dealer—to recognize and evaluate quality and value in cameos, and avoid the pitfalls of purchasing mediocre pieces, fakes and forgeries.
6 x 9, 416 pp, with over 250 photos and illus., 60 in full color; index
Quality Paperback, 978-0-943763-60-6 **$24.99**

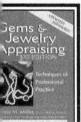

ILLUSTRATED GUIDE TO JEWELRY APPRAISING,
3RD EDITION • *Antique, Period & Modern*
by Anna M. Miller, GG, RMV

This beautifully illustrated guide **provides step-by-step instruction** in jewelry identification and dating, reviews the responsibilities of the appraiser, and describes in detail virtually every style of antique and period jewelry for the hobbyist and serious collector alike.
8½ x 11, 216 pp, with over 150 b/w photos and illus.; index
Hardcover, 978-0-943763-42-2 **$49.99**

GEMS & JEWELRY APPRAISING, 3RD EDITION
Techniques of Professional Practice
by Anna M. Miller, GG, RMV; *revised, updated and expanded by*
Gail Brett Levine, GG

The **premier guide to the standards, procedures and ethics of apprais-ing gems, jewelry and other valuables.** *Gems & Jewelry Appraising* offers all the information that jewelers, gemologists and students will need to establish an appraisal business, handle various kinds of appraisals and provide an accurate, verifiable estimate of value.
8½ x 11, 256 pp, with over 150 b/w photos and illus.; index
Hardcover, 978-0-943763-53-8 **$39.99**

TREASURE HUNTER'S GEM & MINERAL GUIDES TO THE USA
5TH EDITION

Where & How to Dig, Pan and Mine Your Own Gems & Minerals
—IN 4 REGIONAL VOLUMES—
by Kathy J. Rygle *and* Stephen F. Pedersen • *Preface by*
Antoinette Matlins, PG, FGA, *author of* Gem Identification Made Easy

From rubies, opals and gold, to emeralds, aquamarine and diamonds, each guide offers **state-by-state details on more than 250 gems and minerals** and the affordable "fee dig" sites where they can be found. Each guide covers:

• **Equipment & Clothing:** What you need and where to find it.
• **Mining Techniques:** Step-by-step instructions.
• **Gem and Mineral Sites:** Directions & maps, hours, fees and more.
• **Museums and Mine Tours**

All guides: 6 x 9, Quality Paperback Original, illus., maps & photos; indexes.
$14.99 each

Northeast (CT, DC, DE, IL, IN, MA, MD, ME, MI, NH, NJ, NY, OH, PA, RI, VT, WI)
240 pp, 978-0-943763-76-7
Northwest (AK, IA, ID, MN, MT, ND, NE, OR, SD, WA, WY)
200 pp, 978-0-943763-74-3
Southeast (AL, AR, FL, GA, KY, LA, MO, MS, NC, SC, TN, VA, WV)
224 pp, 978-0-943763-77-4
Southwest (AZ, CA, CO, HI, KS, NM, NV, OK, TX, UT)
224 pp, 978-0-943763-75-0

Buy Your "Tools of the Trade ..."

Gem Identification Instruments Directly from *GemStone Press*

Whatever instrument you need, GemStone Press can help.
Use our convenient order form, or contact us directly for assistance.

Complete Pocket Instrument Set
SPECIAL SAVINGS!
BUY THIS ESSENTIAL TRIO AND SAVE 12%
Used together, you can identify 85% of all gems with these three
portable, pocket-sized instruments—the essential trio.
10X Triplet Loupe • Calcite Dichroscope • Chelsea Filter

Pocket Instrument Set:
With Bausch & Lomb 10X Loupe • EZVIEW Dichroscope • Chelsea Filter   **only $179.95**

ITEM / QUANTITY	PRICE EA.*	TOTAL $
Pocket Instrument Set		
_____ With Bausch & Lomb 10X Loupe • EZview Dichroscope • Chelsea Filter	$179.95	_____
Loupes—Professional Jeweler's 10X Triplet Loupes		
_____ Bausch & Lomb 10X Triplet Loupe	$44.00	_____
_____ Standard 10X Triplet Loupe	$29.00	_____
_____ Dark-field Loupe	$58.95	_____
• Spot filled diamonds, identify inclusions in colored gemstones. Operates with Standard Mini Maglite (additional—see below).		
Calcite Dichroscope		
_____ Dichroscope (EZview)	$115.00	_____
Color Filters		
_____ Chelsea Filter	$44.95	_____
_____ Synthetic Emerald Filter Set (Hanneman)	$32.00	_____
_____ Tanzanite Filter (Hanneman)	$28.00	_____
_____ Bead Buyer's & Parcel Picker's Filter Set (Hanneman)	$24.00	_____
Diamond Testers and Tweezers		
_____ SSEF Blue Diamond Tester	$695.00	_____
_____ SSEF Diamond-Type Spotter	$150.00	_____
_____ DiamondNite Dual Tester	$269.00	_____
_____ Diamond Tweezers/Locking	$10.65	_____
_____ Diamond Tweezers/Non-Locking	$7.80	_____
Jewelry Cleaner		
_____ Speed Brite Ionic Jewelry Cleaner	$85.00	_____
_____ Ionic Solution—32 oz. bottle	$22.00	_____

Buy Your "Tools of the Trade ..."

Gem Identification Instruments Directly from *GemStone Press*

Whatever instrument you need, GemStone Press can help.
Use our convenient order form, or contact us directly for assistance.

ITEM / QUANTITY	PRICE EA.*	TOTAL $
Lamps—Ultraviolet & High Intensity		
_____ UV-Blocking Goggles	$24.95	_____
• Recommended for use with all UV lamps.		
_____ Small Longwave/Shortwave (UVP)	$85.00	_____
_____ Large Longwave/Shortwave (UVP)	$237.00	_____
_____ Viewing Cabinet for Large Lamp (UVP)	$195.00	_____
_____ **Purchase Large Lamp & Cabinet together**	$385.95	_____
and save over $45.00		
_____ SSEF High-Intensity Shortwave Illuminator	$499.00	_____
• Operates with SSEF Diamond-Type Spotter (additional—see above).		
Other Light Sources		
_____ Standard Mini Maglite	$15.00	_____
_____ Flex Light	$29.95	_____
Refractometer		
_____ Refractometer	$650.00	_____
_____ Refractive Index Liquid 1.81—10 grams	$69.95	_____
Scale		
_____ GemPro500 Precision Scale	$225.00	_____
Spectroscopes		
_____ Spectroscope—Pocket-sized model (OPL)	$98.00	_____
_____ Spectroscope—Desk model w/stand (OPL)	$235.00	_____

Shipping/Insurance per order in the U.S.: $7.95 first item, SHIPPING/INS. $_____
$3.00 each add'l item; $10.95 total for pocket instrument set.

Outside the U.S.: Please specify *insured* shipping method you prefer
and provide a credit card number for payment. **TOTAL $ _____** **

Check enclosed for $ _____ (Payable to: GEMSTONE PRESS)
Charge my credit card: ❏ Visa ❏ MasterCard
Name on Card _____ Phone (_____)_____
Cardholder Address: Street _____
City/State/Zip _____ E-mail _____
Credit Card # _____ Exp. Date _____
Signature _____ CID # _____
Please send to: ❏ Same as Above ❏ Address Below
Name _____
Street _____
City/State/Zip _____ Phone (_____)_____

Phone, mail, fax, or e-mail orders to:
GEMSTONE PRESS, P.O. Box 237, Woodstock, VT 05091
Tel: (802) 457-4000 • Fax: (802) 457-4004
Credit Card Orders: (800) 962-4544 (8:30AM–5:30PM EST Monday–Friday)
sales@gemstonepress.com • www.gemstonepress.com
Generous Discounts on Quantity Orders

TOTAL SATISFACTION GUARANTEE
If for any reason you're not completely delighted
with your purchase, return it in resellable condition
within 30 days for a full refund.

*Prices, manufacturing specifications and terms subject to change
without notice. Orders accepted subject to availability.

**All orders must be prepaid by credit card, money order or
check in U.S. funds drawn on a U.S. bank.

Please send me:

CAMEOS OLD & NEW, 4TH EDITION
_____ copies at $24.99 (Quality Paperback) *plus s/h**

COLORED GEMSTONES, 3RD EDITION: THE ANTOINETTE MATLINS BUYING GUIDE
_____ copies at $18.99 (Quality Paperback) *plus s/h**

DIAMONDS, 3RD EDITION: THE ANTOINETTE MATLINS BUYING GUIDE
_____ copies at $18.99 (Quality Paperback) *plus s/h**

ENGAGEMENT & WEDDING RINGS, 3RD EDITION: THE DEFINITIVE BUYING GUIDE
_____ copies at $18.95 (Quality Paperback) *plus s/h**

**GEM IDENTIFICATION MADE EASY, 5TH EDITION:
A HANDS-ON GUIDE TO MORE CONFIDENT BUYING & SELLING**
_____ copies at $38.99 (Hardcover) *plus s/h**

GEMS & JEWELRY APPRAISING, 3RD EDITION
_____ copies at $39.99 (Hardcover) *plus s/h**

ILLUSTRATED GUIDE TO JEWELRY APPRAISING, 3RD EDITION
_____ copies at $49.99 (Hardcover) *plus s/h**

**JEWELRY & GEMS AT AUCTION: THE DEFINITIVE GUIDE TO BUYING & SELLING
AT THE AUCTION HOUSE & ON INTERNET AUCTION SITES**
_____ copies at $19.95 (Quality Paperback) *plus s/h**

JEWELRY & GEMS, 7TH EDITION: THE BUYING GUIDE
_____ copies at $19.99 (Quality Paperback) *plus s/h**

THE PEARL BOOK, 4TH EDITION: THE DEFINITIVE BUYING GUIDE
_____ copies at $19.99 (Quality Paperback) *plus s/h**

THE SPECTROSCOPE AND GEMMOLOGY
_____ copies at $49.95 (Quality Paperback) *plus s/h**

**TREASURE HUNTER'S GEM & MINERAL GUIDES TO THE U.S.A., 5TH EDITION:
WHERE & HOW TO DIG, PAN AND MINE YOUR OWN GEMS & MINERALS
IN 4 REGIONAL VOLUMES** $14.99 per copy (Quality Paperback) *plus s/h**
____ copies of NE States ____ copies of SE States ____ copies of NW States ____ copies of SW States

* In U.S.: Shipping/Handling: $3.95 for 1st book, $2.00 each additional book.
 Outside U.S.: Specify shipping method (insured) and provide a credit card number for payment.

Check enclosed for $_____ (Payable to: GEMSTONE Press)

Charge my credit card: ❑ Visa ❑ MasterCard

Name on Card (PRINT) _____ Phone (____)_____

Cardholder Address: Street _____

City / State / Zip _____ E-mail _____

Credit Card # _____ Exp. Date _____

Signature _____ CID # _____

Please send to: ❑ Same as Above ❑ Address Below

Name (PRINT) _____

Street _____

City / State / Zip _____ Phone (____)_____

| **TOTAL SATISFACTION GUARANTEE**
If for any reason you're not completely delighted with your purchase, return it in resellable condition within 30 days for a full refund. | *Phone, mail, fax, or e-mail orders to:*
GEMSTONE PRESS, Sunset Farm Offices,
Rte. 4, P.O. Box 237, Woodstock, VT 05091
Tel: **(802) 457-4000** • *Fax:* **(802) 457-4004**
Credit Card Orders: **(800) 962-4544**
(8:30AM–5:30PM EST Monday–Friday)
sales@gemstonepress.com • www.gemstonepress.com
Generous Discounts on Quantity Orders | Prices subject
to change |

Try Your Bookstore First

Cut along dotted lines and remove, fold as shown at center, tape closed and mail to GemStone Press.

WIN A
$100
GIFT CERTIFICATE!

Fill in this card and
mail it to us—
or fill it in online at
**gemstonepress.com/
feedback.html**
—to be eligible for a
$100 gift certificate for
GemStone Press books and
gem identification equipment.

Place
Stamp
Here

**GEMSTONE PRESS
SUNSET FARM OFFICES RTE 4
PO BOX 237
WOODSTOCK VT 05091-0237**

||ııı.|ı|ı|||ıı.|ı|ıııı|||ıııı.|ı|ıı||ı|.ıı|ıı||ı|ıı|ıı||ı|

_____
(fold here)

**Fill in this card and return it to us to be eligible for our
quarterly drawing for a $100 gift certificate for GemStone Press books.**

We hope that you find this book to be a useful and valuable resource.

Book title: _____

Your comments: _____
_____
_____

How you learned of this book: _____

If purchased: Bookseller _____ City _____ State _____

Please send me a free GemStone Press Publishing catalog. I am interested in: (check all that apply)

Name (PRINT) _____

  1. ❑ Diamonds   2. ❑ Colored Stones   3. ❑ Pearls
  4. ❑ Seminars/Workshops   5. ❑ Gem Identification Equipment

Street _____

City _____ State _____ Zip _____

E-MAIL (FOR SPECIAL OFFERS ONLY) _____

Please send a SkyLight Paths Publishing catalog to my friend:

Name (PRINT) _____

Street _____

City _____ State _____ Zip _____

GemStone Press Tel: (802) 457-4000 • Fax: (802) 457-4004

Available at better booksellers. Visit us online at www.gemstonepress.com

Printed in the USA
CPSIA information can be obtained
at www.ICGtesting.com
JSHW021436221024
72172JS00002B/21

9 780943 763521